INNOVATIONS IN FEDERAL STATISTICS

Combining Data Sources While Protecting Privacy

Panel on Improving Federal Statistics for
Policy and Social Science Research Using Multiple Data Sources and
State-of-the-Art Estimation Methods

Robert M. Groves and Brian A. Harris-Kojetin, *Editors*

Committee on National Statistics

Division of Behavioral and Social Sciences and Education

A Report of

The National Academies of
SCIENCES · ENGINEERING · MEDICINE

THE NATIONAL ACADEMIES PRESS
Washington, DC
www.nap.edu

THE NATIONAL ACADEMIES PRESS 500 Fifth Street, NW Washington, DC 20001

This activity was supported by a grant from the Laura and John Arnold Foundation with additional support from the National Academy of Sciences Kellogg Fund. Support for the work of the Committee on National Statistics is provided by a consortium of federal agencies through a grant from the National Science Foundation (award number SES-1024012). Any opinions, findings, conclusions, or recommendations expressed in this publication do not necessarily reflect the views of any organization or agency that provided support for the project.

International Standard Book Number-13: 978-0-309-45428-5
International Standard Book Number-10: 0-309-45428-X
Digital Object Identifier: 10.17226/24652

Additional copies of this publication are available for sale from the National Academies Press, 500 Fifth Street, NW, Keck 360, Washington, DC 20001; (800) 624-6242 or (202) 334-3313; http://www.nap.edu.

Printed in the United States of America

Suggested citation: National Academies of Sciences, Engineering, and Medicine. (2017). *Innovations in Federal Statistics: Combining Data Sources While Protecting Privacy*. Washington, DC: The National Academies Press. doi: 10.17226/24652.

The National Academies of

SCIENCES • ENGINEERING • MEDICINE

The **National Academy of Sciences** was established in 1863 by an Act of Congress, signed by President Lincoln, as a private, nongovernmental institution to advise the nation on issues related to science and technology. Members are elected by their peers for outstanding contributions to research. Dr. Marcia McNutt is president.

The **National Academy of Engineering** was established in 1964 under the charter of the National Academy of Sciences to bring the practices of engineering to advising the nation. Members are elected by their peers for extraordinary contributions to engineering. Dr. C. D. Mote, Jr., is president.

The **National Academy of Medicine** (formerly the Institute of Medicine) was established in 1970 under the charter of the National Academy of Sciences to advise the nation on medical and health issues. Members are elected by their peers for distinguished contributions to medicine and health. Dr. Victor J. Dzau is president.

The three Academies work together as the **National Academies of Sciences, Engineering, and Medicine** to provide independent, objective analysis and advice to the nation and conduct other activities to solve complex problems and inform public policy decisions. The National Academies also encourage education and research, recognize outstanding contributions to knowledge, and increase public understanding in matters of science, engineering, and medicine.

Learn more about the National Academies of Sciences, Engineering, and Medicine at **www.national-academies.org**.

The National Academies of
SCIENCES • ENGINEERING • MEDICINE

Reports document the evidence-based consensus of an authoring committee of experts. Reports typically include findings, conclusions, and recommendations based on information gathered by the committee and committee deliberations. Reports are peer reviewed and are approved by the National Academies of Sciences, Engineering, and Medicine.

Proceedings chronicle the presentations and discussions at a workshop, symposium, or other convening event. The statements and opinions contained in proceedings are those of the participants and are not necessarily endorsed by other participants, the planning committee, or the National Academies of Sciences, Engineering, and Medicine.

For information about other products and activities of the National Academies, please visit nationalacademies.org/whatwedo.

PANEL ON IMPROVING FEDERAL STATISTICS FOR POLICY AND SOCIAL SCIENCE RESEARCH USING MULTIPLE DATA SOURCES AND STATE-OF-THE-ART ESTIMATION METHODS

ROBERT M. GROVES (*Chair*), Provost, Georgetown University
MICHAEL E. CHERNEW, Department of Health Care Policy, Harvard Medical School
PIET DAAS, Department of Corporate Services, Information Technology, and Methodology, Statistics Netherlands
CYNTHIA DWORK, John A. Paulson School of Engineering and Applied Sciences and Radcliffe Institute for Advanced Study, Harvard University
OPHIR FRIEDER, Department of Computer Sciences, Georgetown University
HOSAGRAHAR V. JAGADISH, Computer Science and Engineering, University of Michigan
FRAUKE KREUTER, Joint Program in Survey Methodology, University of Maryland, and Statistics and Methodology, University of Mannheim and Institute for Employment Research
SHARON LOHR, Vice President, Westat, Rockville, Maryland
JAMES P. LYNCH, Department of Criminology and Criminal Justice, University of Maryland
COLM O'MUIRCHEARTAIGH, Harris School of Public Policy Studies, University of Chicago
TRIVELLORE RAGHUNATHAN, Institute for Social Research, University of Michigan
ROBERTO RIGOBON, Sloan School of Management, Massachusetts Institute of Technology
MARC ROTENBERG, President, Electronic Privacy Information Center, Washington, DC

BRIAN HARRIS-KOJETIN, *Study Director*
HERMANN HABERMANN, *Senior Program Officer*
GEORGE SCHOEFFEL, *Research Assistant*
AGNES GASKIN, *Administrative Assistant*

COMMITTEE ON NATIONAL STATISTICS

Acknowledgments

This report of the Panel on Improving Federal Statistics for Policy and Social Science Research Using Multiple Data Sources and State-of-the-Art Estimation Methods is the product of contributions from many colleagues, whom we thank for their generous sharing of their time and expertise.

The panel is grateful to the Laura and John Arnold Foundation for funding this study, and to foundation staff Stuart Buck and Meredith McPhail for their help and guidance throughout the study. The panel also is grateful for the supplemental funding provided by the National Academy of Sciences Kellogg Fund.

The panel thanks Katherine Wallman, recently retired chief statistician at the U.S. Office of Management and Budget, and the heads of the principal statistical agencies for their valuable insights: Mary Bohman, Economic Research Service; Peggy Carr, National Center for Education Statistics; John R. Gawalt, National Center for Science and Engineering Statistics; Erica L. Groshen, Bureau of Labor Statistics; Hubert Hamer, National Agricultural Statistics Service; Patricia Hu, Bureau of Transportation Statistics; Barry Johnson, Statistics of Income Division of the Internal Revenue Service; Brian Moyer, Bureau of Economic Analysis; Jeri Mulrow, Bureau of Justice Statistics; John W.R. Phillips, Office of Research, Evaluation, and Statistics in the Social Security Administration; Charles J. Rothwell, National Center for Health Statistics; Adam Sieminski, Energy Information Administration; and John H. Thompson, U.S. Census Bureau. Their contributions and support to the panel during our initial meeting as well as their support and encouragement throughout the study have been invaluable in helping the panel understand the challenges and constraints that the federal statistical agencies face and their dedication to providing high-quality information for the public good.

The panel also thanks all the many individuals who participated in one or more of the panel's three workshops. A list of the presenters at the workshops can be found in Appendix A. The panel also thanks Steve Eglash (Stanford University) for his work examining issues of data access for private-sector companies.

At the National Academies of Sciences, Engineering, and Medicine, the panel would not have been able to complete its work efficiently without a capable staff. Connie Citro, director of the Committee on National Statistics, had the vision and perseverance to make this study a reality. Mary Ellen O'Connell, director of the Division of Behavioral and Social Sciences and Education, and Robert Hauser, previous director of the division, provided both institutional leadership and substantive insights. The division's Kirsten Sampson-Snyder was extremely helpful in coordinating the review process, and Eugenia Grohman provided meticulous and thorough editing that greatly improved the readability of the report.

For the Committee on National Statistics, both Agnes Gaskin, administrative assistant, and Anthony Mann, program coordinator, provided considerable assistance in managing the logistics of this panel and their meetings in various geographic and institutional locations. Hermann Habermann, senior program officer, provided valuable feedback and guidance on drafts of this report, and George Schoeffel, research assistant, cheerfully assisted with every aspect of the study, performing countless tasks to make this report possible. Most critically, Brian Harris-Kojetin served as study director and not only kept the panel focused on the relevant tasks at hand, but also provided much expertise in the report and responded to the many comments and questions from panel members and reviewers. Without his technical skill, organizational skill, and insight, this report would not nearly be what it is currently.

This report has been reviewed in draft form by individuals chosen for their diverse perspectives and technical expertise. The purpose of this independent review is to provide candid and critical comments that will assist the institution in making the published report as sound as possible and to ensure that the report meets the institutional standards for objectivity, evidence, and responsiveness to the study charge. The review comments and draft manuscript remain confidential to protect the integrity of the deliberative process.

We thank the following individuals for their participation in the review of the report: John M. Abowd, research and methodology, U.S. Census Bureau; Cynthia Z.F. Clark, independent consultant, McLean, Virginia; Arthur B. Kennickell, research and statistics, Board of Governors of the

Federal Reserve System; Partha Lahiri, Joint Program in Survey Methodology, University of Maryland; Thomas A. Louis, Department of Biostatistics, Bloomberg School of Public Health, Johns Hopkins University; Nancy Mathiowetz, Department of Sociology (emerita), University of Wisconsin–Milwaukee; Thomas L. Mesenbourg, U.S. Census Bureau (retired); and Alan M. Zaslavsky, Department of Health Care Policy, Harvard Medical School.

Although the reviewers listed above provided many constructive comments and suggestions, they were not asked to endorse the report's conclusions or recommendations, nor did they see the final draft of the report before its release. The review of this report was overseen by Michael Hout, Department of Sociology, New York University, and Alicia L. Carriquiry, Department of Statistics and Sciences, Iowa State University. They were responsible for making certain that an independent examination of this report was carried out in accordance with institutional procedures and that all review comments were carefully considered. Responsibility for the final content of this report rests entirely with the authoring panel and institution.

Robert M. Groves, *Chair*
Panel on Improving Federal Statistics for Policy and Social Science Research Using Multiple Data Sources and State-of-the-Art Estimation Methods

Contents

Executive Summary

Federal government statistics provide critical information to the country and serve a key role in a democracy. For decades, sample surveys with instruments carefully designed for particular data needs have been one of the primary methods for collecting data for federal statistics. However, the costs of conducting such surveys have been increasing while response rates have been declining, and many surveys are not able to fulfill growing demands for more timely information and for more detailed information at state and local levels.

The Panel on Improving Federal Statistics for Policy and Social Science Research Using Multiple Data Sources and State-of-the-Art Estimation Methods was charged to conduct a study to foster a paradigm shift in federal statistical programs that would use combinations of diverse data sources from government and private-sector sources in place of a single census, survey, or administrative records source. This first report discusses the challenges faced by the federal statistical system and the foundational elements needed for a new paradigm.

In addition to surveys, some federal statistics are also derived from government administrative records, that is, data collected by government entities for program administration, regulatory, or law enforcement purposes. Because these administrative records exist, there is interest in using them much more—both alone and in combination with surveys—to try to enhance the quality, scope, and cost-efficiency of statistical products and to reduce response burden on the public.

Not enough is known about the quality of these new sources of data, and considerable work is required to assess their usefulness for producing statistics. Some may be useful as is; other may require scrubbing or statistical transformation. Furthermore, for statistical purposes, it may be necessary to combine or blend multiple data sources, which is more complex than working with a single dataset. However, there are statistical methods and models for combining information from multiple data sources.

Some administrative records held by federal agencies are prohibited from being shared among agencies. And for some records held by states and localities, there is no mandate and limited incentive to share them with federal statistical agencies.

> **CONCLUSION 3-4** Legal and administrative barriers limit the statistical use of administrative datasets by federal statistical agencies.

> **CONCLUSION 3-5** State and local governments may respond to incentives from the federal government to provide access to their administrative data by federal statistical agencies for statistical purposes.

> **RECOMMENDATION 3-1** Federal statistical agencies should systematically review their statistical portfolios and evaluate the potential benefits and risks of using administrative data. To this end, federal statistical agencies should create collaborative research programs to address the many challenges in using administrative data for federal statistics.

Large amounts of private-sector data—such as credit card transactions, scanner data, cell phone data, and Internet searches—are generated for commercial use. These sources hold the potential to improve the timeliness and level of detail of national statistics. These data are extremely diverse, and there are many issues of access, quality, and usability that would have to be addressed to consider them for federal statistical use.

> **RECOMMENDATION 4-1** Federal statistical agencies should systematically review their statistical portfolios and evaluate the potential benefits of using private-sector data sources.

> **RECOMMENDATION 4-2** The Federal Interagency Council on Statistical Policy should urge the study of private-sector data and evaluate both their potential to enhance the quality of statistical products and the risks of their use. Federal statistical agencies should provide annual public reports of these activities.

Any consideration of expanding the use of data must have privacy as a core value. Federal privacy laws have established clear limitations on the collection and use of personally identifiable information, and statistical agencies have a strong tradition of data confidentiality and stewardship. Nonetheless, data breaches pose real risks to the public. As federal statistical agencies seek to combine multiple datasets, they need to simultaneously address how to control risks from privacy breaches. Privacy-enhancing techniques and privacy-preserving statistical data analysis can be valuable in these efforts and enable the use of private-sector and other alternative data sources for federal statistics.

> **RECOMMENDATION 5-1** Statistical agencies should engage in collaborative research with academia and industry to continuously develop new techniques to address potential breaches of the confidentiality of their data.

> **RECOMMENDATION 5-2** Federal statistical agencies should adopt modern database, cryptography, privacy-preserving, and privacy-enhancing technologies.

In the decentralized U.S. statistical system, there are 13 agencies whose mission is primarily the creation and dissemination of statistics and more than 100 agencies that engage in statistical activities. However, there is currently no agency directly charged with facilitating access to and the use of multiple data sources for the benefit of the entire statistical system. There is a need for stronger coordination and collaboration to enable access to and evaluation of administrative and private-sector data sources for federal statistics.

> **RECOMMENDATION 6-1** A new entity or an existing entity should be designated to facilitate secure access to data for statistical purposes to enhance the quality of federal statistics.

Privacy protections would have to be fundamental to the mission of this entity.

> **CONCLUSION 6-1** For the proposed new entity to be sustainable, the data for which it has responsibility would need to have legal protections for confidentiality and be protected, using the strongest privacy protocols offered to personally identifiable information while permitting statistical use.

RECOMMENDATION 6-2 The proposed new entity should maximize the utility of the data for which it is responsible while protecting privacy by using modern database, cryptography, privacy-preserving, and privacy-enhancing technologies.

There are many questions about how the entity would function and who would be able to access data for statistical purposes. The panel's second report will examine organizational models for a new entity, quality frameworks for multiple data sources, statistical techniques for combining data from multiple sources, privacy-enhancing and privacy-preserving techniques, as well as the information technology implications for implementing a new paradigm that would combine diverse data sources.

1

Introduction

At 8:30 a.m. on the first Friday of every month, the Bureau of Labor Statistics (BLS) announces the employment situation for the United States, which includes the count of new jobs and the unemployment rate. These statistics are scrutinized by economists, policy makers, and advocacy groups, and they influence a broad range of decisions by governments, businesses, and the general public. The monthly announcement can result in the movement of more than $150 billion in investments in the U.S. stock markets within minutes of release (e.g., see Saslow, 2012).

Other federal statistics are similarly influential. They are used in allocation formulas that direct the annual flow of more than $400 billion in federal funds to state and local governments for a wide variety of programs and purposes (Blumerman and Vidal, 2009; National Research Council, 2003; Reamer and Carpenter, 2010; U.S. Government Accountability Office, 2009a, 2009b). Statistics on consumer prices are used to adjust tax rates and government program benefits, such as Social Security, for cost-of-living increases.[1] Whether people realize it or not, federal statistics continuously touch their lives.

Historically, the primary vehicle for statistical agencies to collect useful information has been sample surveys, administered to individuals, households, farms, businesses, governments, schools, health care providers, and others. These surveys are based on well-accepted principles of statistical sampling designed to produce a representative group of respondents. The BLS estimate of the total number of new jobs each month comes from a

[1]See http://www.bls.gov/cpi/cpifaq.htm#Question_8 [December 2016].

survey that covers more than 600,000 business establishments (Current Employment Statistics program). The unemployment rate comes from a survey of more than 50,000 households (the Current Population Survey). Of wider significance, federal surveys have contributed to important public policy initiatives and new social science knowledge in fields as varied as science and engineering resources, agricultural output, assistance for low-income people, crime victimization, housing quality, business ownership, health care costs and quality, educational attainment, labor force experience, and how people use their time and feel about their lives.

Despite their importance, the sustainability of many federal surveys is threatened by declining response rates and increased costs for data collection. Yet at the same time that statistical agencies have been facing flat or decreasing budgets, they are facing growing demands by the business community and state and local governments for more geographically detailed and more timely data. The rest of this chapter first describes the important role of federal statistics and notes important parallels with current initiatives on program evaluation and evidence-based policy making and then details the charge to the panel and our activities. We end with a brief overview of the report.

FEDERAL STATISTICS AND EVIDENCE-BASED POLICY MAKING

Many federal statistical products, such as those noted above, are labeled as "descriptive." They answer such questions as "how much?" (as in the number of jobs created in a month) or "how prevalent?" (as in the percentage of adults in the labor force).

However, key statistics produced by federal statistical agencies and the underlying survey data provided by these agencies also provide a vital information infrastructure to inform and evaluate public policies. Indeed, many researchers rely on survey data from federal statistical agencies as one important source for policy analysis and other social science research to examine critical social and economic issues.

In contrast to the descriptive uses of data, these uses of data are sometimes referred to as "analytic" or "research based." Analytic statistics and research uses of the data often focus on the "how" and "why" of various outcomes. Are the higher incomes of job-training participants (compared with those who did not receive job training) the result of the training or some other aspect or change in their lives? Evidence-based policy making requires answering questions about whether government programs produce their desired effects.

Evaluations of programs are designed to identify the mechanisms in a program that are most important to achieve the desired effects. The better the design of such studies, the more effectively the mechanisms can be identified. In fact, given the nature of social science, broader access, by different research groups, is often needed to reach a consensus on the effects of existing programs or to project the potential effects of proposed programs.

Assessing the effectiveness of a program can often be based on the same data that are used in federal statistical agencies to produce descriptive statistics. Program administrative data may also be used to examine the outcomes of participants at one time or to follow them over time to examine longer term outcomes. One might examine the later employment and wages of participants in a job training or education program with data from their tax records to assess how effective that program was. These evaluation studies are often carried out by federal contractors or academic researchers, who formulate the research questions, determine the measures, collect or acquire the appropriate data, analyze the data, and publish the results.

There has been increasing attention in recent years to evidence-based policy making, which can use a variety of data sources, research methodologies, and analytic approaches. The Obama administration asked Congress for resources to build evaluation capacity within agencies and expand infrastructure for researchers to have access to federal survey and administrative data for evaluation studies (U.S. Office of Management and Budget, 2015a, 2016). As noted above, those data are collected by government entities for program administration, regulatory, or law enforcement purposes, and they include such records as employment and earnings information on state unemployment insurance records, income reported on federal tax forms, Social Security earnings and benefits, medical conditions and payments made for services from Medicare and Medicaid records, and food assistance program benefits (see U.S. Office of Management and Budget, 2014a). In 2016, Congress established an Evidence-Based Policymaking Commission, which will examine arrangements for integrating federal survey and administrative data and making those data available to researchers (P.L. 114-140).

Federal statistical agencies could also benefit from improved access to administrative and other data sources. There are many potentially valuable nonsurvey data sources—such as federal, state, and local government administrative records, private-sector credit card transactions, sensor data, geospatial data—and a wide and growing variety of web-based data, such as text and images from social media sites. These data have the potential to provide significant improvements to federal statistical programs, which often now rely only on survey-based datasets, in timeliness, geographic detail, and cost-effectiveness. To the extent that the use of other data sources makes it possible to enrich the quality of federal statistics without increasing (or perhaps even decreasing) the burden on survey respondents, the federal statistical system can more efficiently serve the country.

Making greater use of other data sources for federal statistics is also important because of declining survey response rates (Czajka and Beyler,

2016; National Research Council, 2013a), high and increasing nonresponse to key items, such as income (Czajka and Denmead, 2008; Meyer et al., 2015), and rising per-unit costs. Indeed, the problem was clearly described in a study by the National Research Council (2013a, p. 7):

> Household survey responses rates in the United States have been steadily declining for at least the last two decades. A similar decline in survey response can be observed in all wealthy countries, and is particularly high in areas with large numbers of single-parent households, families with young children, workers with long commutes, and high crime rates. Efforts to raise response rates have used monetary incentives or repetitive attempts to obtain completed interviews, but these strategies increase the costs of surveys and are often unsuccessful.

Using new data sources in combination with surveys is not without risk, and there are many challenges with access to these potential new data sources. They will also require both careful evaluations of quality and fitness for specific uses in federal statistics and careful implementation, taking into consideration the importance of the continuity of long-running statistical series. These efforts need to be initiated as soon as possible because they will take time, resources, and collaborative research among agencies and with academia and industry.

PANEL CHARGE AND ACTIVITIES

The Committee on National Statistics (CNSTAT), in the Division of Behavioral and Social Sciences and Education (DBASSE) at the National Academy of Sciences, Engineering, and Medicine, received funding from the Laura and John Arnold Foundation to convene an ad hoc committee of nationally renowned experts in social science research, sociology, survey methodology, economics, statistics, privacy, public policy, and computer science to foster a possible shift in federal statistical programs—from the current approach of providing users with the output from a single census, survey, or administrative records source to a new paradigm of combining data sources with state-of-the-art methods. The goal of such a shift would be to give users richer and more reliable datasets that lead to new insights about policy and socioeconomic behavior. The statement of task for the panel is shown in Box 1-1.

As detailed in the statement of task, this first report of this panel reviews the current approach for producing federal statistics, examines other data sources that could also be used for federal statistics, and discusses the environment needed for using multiple data sources in the future, including statistical methods of combining data sources, mechanisms for research access, and approaches for protecting privacy and preserving confidentiality. A second report will focus on implementation of a new approach for producing federal statistics from multiple data sources including evaluating quality metrics, statistical models for combining data, and methods for preserving privacy. It will also provide recommendations for needed research to move forward with a paradigm of using multiple data sources for federal statistics.

As part of its fact-gathering activities, the panel sponsored three public workshops (see Appendix A for the workshop agendas).[2] In addition, prior to the workshops, the panel held an open session in September 2015, which included a discussion with 10 of the heads of the 13 principal statistical agencies. This session informed the panel about current challenges in day-to-day operations, the challenges in approaching innovation and change, and concerns about the future of the agencies' work. The discussions also informed the panel about the current practices of the statistical agencies and their future plans to deal with these challenges.

The panel's first workshop, held in December 2015, explored how federal statistical agencies are currently using alternative data sources, including a discussion of issues of how federal statistical agencies are currently able to access and use administrative and other nonsurvey sources of data for national statistics. The workshop included discussion of legal and policy issues in accessing alternative data sources, as well as the efforts of statistical agencies to evaluate both the quality of these alternative data sources and methods for combining multiple data sources. The workshop included 20 speakers from federal statistical agencies who described how they were using alternative data sources, including administrative records, private company data, or other data sources in order to create new products or improve existing statistical programs. The workshop also included a presentation and discussion on public perceptions toward federal statistical agencies' use of administrative records.

The second workshop, held in February 2016, focused on how some private-sector firms are using "big data," such as Internet-based search terms, geolocation data, credit card transactions, and data from social media websites. The workshop explored issues of accessing and using a variety of different kinds of data from private sources as well as the access arrangements and safeguards the private sector uses to protect privacy and confidentiality of data for research uses. The workshop also included a discussion of potential models for sharing data among different organizations and ways to use big data for research and statistical purposes.

---

[2]Copies of the workshop presentations are available at http://sites.nationalacademies.org/DBASSE/CNSTAT/DBASSE_170269 [November 2016].

BOX 1-1
Statement of Task

An ad hoc panel of nationally renowned experts in social science research, computing technology, statistical methods, privacy, and use of alternative data sources in the United States and abroad will conduct a study with the goal of fostering a paradigm shift in federal statistical programs. In place of the current paradigm of providing users with the output from a single census, survey, or administrative records source, a new paradigm would use combinations of diverse data sources from government and private sector sources combined with state-of-the-art methods to give users richer and more reliable statistics leading to new insights about policy and socioeconomic behavior. The motivation for the study stems from the increasing challenges to the current paradigm, such as declining response rates and increasing cost and burden for surveys.

The panel will prepare two reports as part of this study:

First Report

The first report will discuss the challenges faced by the federal statistical system; the current paradigm of providing users with the output from a single census, survey, or administrative records source and that paradigm's increasing disadvantages for meeting user needs; and the foundational elements needed for a new paradigm.

More specifically, the first report will discuss:

- federal statistical agencies' current paradigm for producing national statistics and challenges to this paradigm;
- federal statistical agencies' legal frameworks and mechanisms for protecting the privacy and confidentiality of their data and challenges to those frameworks and mechanisms;
- federal statistical agencies' legal frameworks and mechanisms for providing access to underlying data to researchers to foster transparency, replicability of statistical series, and for policy and social science research and challenges to those frameworks and mechanisms;
- federal statistical agencies' access to alternative sources of data for federal statistical programs, the organizational structures sustaining access, and the impediments to access;

The third workshop, held in June 2016, examined state and local governments' use of administrative and other data sources, including how local integrated data systems are created, governed, and used to improve community services. The workshop also included discussions on some of the difficulties in trying to establish integrated data systems, obtaining and

- the characteristics of a new paradigm for federal statistical programs that would combine diverse data sources from government and private sector sources with state-of-the-art methods to give users richer and more reliable statistics; and
- the foundational elements needed for a new paradigm.

The first report will contain findings and conclusions from the panel's deliberations and recommendations for steps needed to lay the foundation for a new paradigm.

Second Report

The second report will propose approaches for implementing a new paradigm that would combine diverse data sources from government and private sector sources with state-of-the-art methods to give users richer and more reliable statistics.

The second report will:

- assess alternative approaches for implementing a new paradigm that would combine diverse data sources from government and private sector sources;
- evaluate concepts, metrics, and methods for assessing the quality and utility of alternative data sources, analogous to the "total error" framework used for surveys;
- evaluate statistical models for combining data from multiple sources;
- examine metrics and methods for evaluating the quality of combined-information estimates;
- evaluate alternative designs of statistical processes that foster privacy protections, transparency, objectivity, timeliness, replicability, efficiency, and continuity of statistical series; and
- identify priorities for research needed for federal statistical agencies to advance a multiple-data-sources paradigm.

The second report will contain findings, conclusions, and recommendations for actions toward implementing a new multiple-data-sources paradigm for federal statistics.

maintaining data, and quality issues with various types of data. In addition, the workshop explored the use of sensor data, which can monitor pollution, light, and traffic, as well as privacy issues with using sensor data and ways of designing systems to incorporate privacy.

OVERVIEW OF THE REPORT

The next three chapters discuss the data sources for federal statistics. Chapter 2 is a brief history of the federal statistical system, focusing on the use of sample surveys to produce statistics. Chapter 3 reviews the role of administrative records in the U.S. federal statistical system in comparison with other countries, the benefits of and challenges with these data, and current efforts to make greater use of administrative records for federal statistics. Chapter 4 describes some private-sector data sources that might be usable for federal statistics, the benefits of and challenges with these data, and current efforts in the United States and in the national statistical offices of other countries to explore and use these alternative sources.

The last two chapters begin to lay a foundation for a new approach to federal statistics and social science research. Chapter 5 provides a brief overview of privacy and confidentiality laws and practices for statistical data, as well as the mechanisms for providing access to data for research uses outside the federal statistical agencies. Chapter 6 presents a new approach for federal statistical programs, which would combine survey, administrative, and private-sector data sources to give users richer and more reliable statistics. Key to this approach are privacy protections and increased access to administrative and other data sources for federal statistical programs.

2

Current Challenges and Opportunities in Federal Statistics

This chapter discusses the importance of federal statistics for the country and provides an overview of the federal statistical system, which is responsible for providing relevant, credible, and timely information to inform policy makers and the public. We describe how sample surveys have come to dominate federal statistics, as well as the current threats to this paradigm, including declining response rates and rising costs. We conclude the chapter with a discussion of the growing demand and expectations for more timely information and their implications for trying to continue to rely solely on sample surveys for federal statistics.

THE U.S. FEDERAL STATISTICAL SYSTEM

Since the founding days of the country, the system of national statistics has changed many times in response to growing needs for data, developments in technology and statistical methodology, decreasing response rates, and increasing concern about privacy (Bellhouse, 2000; Duncan, 1976; Sylvester and Lohr, 2005). The U.S. federal statistical system is highly decentralized, with statistical activities spread across approximately 125 agencies of the federal government (U.S. Office of Management and Budget, 2015b). There are 13 principal statistical agencies, whose primary mission is producing statistics.[1] The U.S. Office of Management and Budget (OMB)

[1]These agencies are the Bureau of Economic Analysis; Bureau of Justice Statistics; Bureau of Labor Statistics; Bureau of Transportation Statistics; Census Bureau; Economic Research Service; Energy Information Administration; National Agricultural Statistics Service; National

is charged (44 U.S. Code §3504(e)) with coordinating the federal statistical system, including issuing standards, guidelines, and statistical directives to all agencies to ensure that agencies use common classifications, definitions, and appropriate methodologies in producing statistical products, as well as with enforcing standards through centralized review of all agency information collections as required by the Paperwork Reduction Act (P.L. 104-13).

At OMB, the U.S. chief statistician leads a small staff in the Statistical and Science Policy Branch (SSP) of the Office of Information and Regulatory Affairs (OIRA) to carry out these activities and chairs the Interagency Council on Statistical Policy (ICSP), which is composed of the heads of the principal statistical agencies and is codified in the Paperwork Reduction Act (44 U.S. Code §3504(e)(8)). ICSP improves coordination and communication across the system through monthly meetings to discuss activities and issues across the agencies, to exchange information about agency programs and activities, and to provide advice and counsel to OMB on statistical matters.

Over the past century, there have been a number of studies of the U.S. federal statistical system that have documented needed improvements. Norwood (1995) reviewed 15 different committees, commissions, and study groups that examined the federal statistical system in the 20th century, and she found that these groups uniformly recommended greater centralization of the system or greater coordination of the decentralized system. However, few actions have ever been taken on those recommendations.

THE IMPORTANCE OF FEDERAL STATISTICS

Federal statistics shape decisions by the public, by businesses, and by government agencies, such as the Federal Reserve Board of Governors, which sets monetary policy and the target for the federal funds interest rate. Many federal statistics are eagerly awaited since the business community demands timely information on the economy. The Bureau of Economic Analysis (BEA) publishes three estimates for the U.S. gross domestic product (GDP) for each quarter, with the "initial" estimate released 30 days after the end of each quarter to provide as timely information as possible. The following month a "second" estimate is released: it includes more complete data than was available the previous month. The next month, the "third" estimate for the quarter is released, based on the most complete data.[2] There can be substantial revisions among these estimates, but the

Center for Education Statistics; National Center for Health Statistics; National Center for Science and Engineering Statistics; Office of Research, Evaluation and Statistics in the Social Security Administration; and Statistics of Income Division in the Internal Revenue Service.

[2]See http://www.bea.gov/methodologies/index.htm#national_meth [January 2017].

demand for timely information seems to outweigh concerns about later changes.

Federal statistics, such as GDP and the employment situation noted in Chapter 1, represent 2 of the 36 designated principal federal economic indicators, which OMB recognizes as statistics that have the potential to move markets when publicly released. Therefore, these statistics are subject to careful controls on the timing and handling of these statistical releases (U.S. Office of Management and Budget, 1985).

Beyond the economy, statistical reports of the crime victimization and criminal justice activities are critical for having informed national discussions about policing, sentencing, crime, and race, as well as legislative efforts to reform the criminal justice system. Statistical evidence about the relative prevalence of health conditions is used to allocate funding to ameliorate those conditions. Epidemiological statistics about the prevalence of infectious disease are used to allocate resources to combat epidemics (National Institutes of Health, 2016). In short, federal statistics matter.

Statistics for the common good are embedded in the very foundation of the United States and are central for the nation's democratic foundation and its economic and social well-being. Statistics derived from high-quality data promote informed decision making and strengthen democratic institutions by informing the public and enabling them to hold leaders accountable. The Constitution (Article 1, Section 2) specifies a decennial census to ensure proportional representation in the House of Representatives, beginning with the first census conducted in 1790. The first statistical agency was created in the 1860s in the Treasury Department, followed by the establishment of units in the Departments of Agriculture and Education (see Bureau of the Census, 1975; Norwood, 1995). As needs for information grew, new statistical agencies were formed in various other departments.

The United States is not unique in its belief in the importance of objective statistical information. Almost every country in the world has established a system of statistical indicators that cover macroeconomic performance, health, labor, agriculture, demography, crime, tobacco and drug use, transportation, and energy that assist in planning, investments, and the development of national priorities. Indeed an informed public requires information about the status of the country (Holdren, 2010; Prewitt, 2010). In that sense, statistical information about the welfare of the country is indispensable to a well-functioning democracy (Holt, 2007; Norwood, 2016).

Given the importance of national statistics, their quality and timeliness also matter a great deal. Even a slight underestimate or overestimate can have multibillion dollar impacts on the country. For example, the report of the Boskin Commission (Advisory Commission to Study the Consumer Price Index, 1996) estimated that the consumer price index produced by the

Bureau of Labor Statistics (BLS) at that time overstated the cost of living by 1.1 percentage points and noted that, if true, "would contribute about $148 billion to the deficit in 2006 and $691 billion to the national debt by then" (pp. 1-2).

In another example, Reamer (2014, pp. 3-4) describes how the lack of regularly updated services industries data prior to 2009 affected the quarterly GDP measure at the beginning of the great recession:

> [T]he erroneous January 2009 prediction by Christina Romer and Jared Bernstein, President-elect Obama's top economic advisers, that the passage of a Recovery Plan of "slightly over $775 billion" would keep the national unemployment rate below 8 percent was based on the overly optimistic GDP data available at the time. At the time, the latest available GDP estimates were a 1.0 percent annual rate of growth in the first quarter of 2008 (2008Q1), a 2.8 percent annual growth rate in 2008Q2, and a 0.5 percent annual rate of decline in 2008Q3. By 2011, BEA had revised these numbers to minus 1.8 percent, plus 1.3 percent, and minus 3.7 percent, respectively.
>
> By the time Congress passed the Recovery Act in February 2009 ($787 billion initial estimate), BEA had issued another, relatively dire GDP number, an annual rate of decline in 2008Q4 of minus 3.8 percent. By 2011, BEA revised that figure to minus 8.9 percent.

These examples demonstrate how important it is that the federal statistical system provide the most accurate and timely indicators, as the country's economic well-being relies upon the accuracy of these statistics. Economic and social policy is in large extent informed by measurements that are produced by statistical offices. Therefore, mistakes in the measurement will be translated to mistakes in policies. Such mistakes are invariably costly. Inaccuracies in statistical indicators could mean unfair allocation of funds among states (National Research Council, 2003; Seeskin and Spencer, 2015; U.S. Government Accountability Office, 2003, 2009a) and could also affect governmental and business decision making. For example, mismeasurement of the inflation rate has consequences on wage negotiations, retirement income, and asset prices. Similarly, mismeasurement of economic activity has consequences for fiscal and economic policy decisions.

It is of critical importance for the country not only that federal statistical agencies provide indicators that assist planning, investments, and the development of national priorities, but also that they do so in an objective manner. Since all statistics have limitations, the credibility of statistical information requires transparency of methods (Miller, 2010), documentation of error qualities, and absolute protection from political interference in the production and dissemination of the statistics (see National Research

Council, 2013b). For that reason, strict codes of ethics, laws, and regulations that protect the operations of national statistical offices from political interference, and powerful pledges of confidentiality of data have been promulgated by national and international organizations.

In the United States, the National Research Council first published *Principles and Practices for a Federal Statistical Agency* in 1992, and it has been widely used by U.S. statistical agencies and cited in OMB directives and GAO reports. It has been and continues to be updated every 4 years and is now in its fifth edition (National Research Council, 2013b). Underlying the production of federal statistics and their usefulness are four principles: (1) relevance to policy issues, (2) credibility among data users, (3) trust among data providers, and (4) independence from political and other undue external influences. The publication also delineates 13 practices that agencies should follow to help achieve and embody these principles (see Box 2-1).

Similarly, the General Assembly of the United Nations formally adopted the Fundamental Principles of Official Statistics in January 2014, providing high-level recognition of the principles that had been promul-

BOX 2-1
Principles and Practices for a Federal Statistical Agency*

To implement the four principles promulgated in *Principles and Practices for a Federal Statistical Agency—Fifth Edition* (National Research Council, 2013b) are 13 practices for a federal statistical agency:

1. a clearly defined and well-accepted mission,
2. necessary authority to protect independence,
3. continual development of more useful data,
4. openness about sources and limitations of data provided,
5. wide dissemination of data,
6. cooperation with data users,
7. respect for the privacy and autonomy of data providers,
8. protection of the confidentiality of data providers' information,
9. commitment to quality and professional standards of practice,
10. an active research program,
11. professional advancement of staff,
12. a strong internal and external evaluation program, and
13. coordination and collaboration with other statistical agencies.

*See https://www.nap.edu/catalog/18318/principles-and-practices-for-a-federal-statistical-agency-fifth-edition [January 2017].

BOX 2-2
Fundamental Principles of Official Statistics

The United Nations General Assembly adopted 10 fundamental principles of official statistics on January 29th, 2014.

Principle 1: Official statistics provide an indispensable element in the information system of a democratic society, serving the Government, the economy and the public with data about the economic, demographic, social and environmental situation. To this end, official statistics that meet the test of practical utility are to be compiled and made available on an impartial basis by official statistical agencies to honour citizens' entitlement to public information.

Principle 2: To retain trust in official statistics, the statistical agencies need to decide according to strictly professional considerations, including scientific principles and professional ethics, on the methods and procedures for the collection, processing, storage and presentation of statistical data.

Principle 3: To facilitate a correct interpretation of the data, the statistical agencies are to present information according to scientific standards on the sources, methods and procedures of the statistics.

Principle 4: The statistical agencies are entitled to comment on erroneous interpretation and misuse of statistics.

Principle 5: Data for statistical purposes may be drawn from all types of sources, be they statistical surveys or administrative records. Statistical agencies are to choose the source with regard to quality, timeliness, costs and the burden on respondents.

Principle 6: Individual data collected by statistical agencies for statistical compilation, whether they refer to natural or legal persons, are to be strictly confidential and used exclusively for statistical purposes.

Principle 7: The laws, regulations and measures under which the statistical systems operate are to be made public.

Principle 8: Coordination among statistical agencies within countries is essential to achieve consistency and efficiency in the statistical system.

Principle 9: The use by statistical agencies in each country of international concepts, classifications and methods promotes the consistency and efficiency of statistical systems at all official levels.

Principle 10: Bilateral and multilateral cooperation in statistics contributes to the improvement of systems of official statistics in all countries.

gated by the U.N. Statistical Commission since 1994 (see Box 2-2). OMB also issued Statistical Policy Directive No. 1: Fundamental Responsibilities of Federal Statistical Agencies and Recognized Statistical Units (Office of Management and Budget, 2014b) (see Box 2-3). This directive enumerates the responsibilities of federal statistical agencies in the design,

BOX 2-3
OMB Statistical Policy Directive No. 1: Fundamental Responsibilities of Federal Statistical Agencies and Recognized Statistical Units

The U.S. Office of Management and Budget Directive No. 1, Fundamental Responsibilities of Federal Statistical Agencies and Recognized Statistical Units, outlines four fundamental responsibilities for the Federal Statistical system:

1. producing and disseminating relevant and timely information,
2. conducting credible and accurate statistical activities,
3. conducting objective statistical activities, and
4. protecting the trust of information providers by ensuring the confidentiality, and exclusive statistical use of their responses.

collection, processing, editing, compilation, storage, analysis, release, and dissemination of statistical information.

Combinations of political pressures, inaccurate data collection, and other poor practices can have severe consequences for both statistical offices and an entire country, as was seen recently in Greece and Argentina (Hartman et al., 2014; *The Economist*, 2012). At the same time, unsubstantiated political attacks on the quality of statistics can harm the credibility of not only particular statistics, but also the agency producing the statistics. Jack Welch, former chief executive officer of General Electric, sent a tweet accusing BLS of "cooking the books" on its release of the unemployment rate on the eve of the 2012 presidential election (Malone and Mutikani, 2012). After gaining some initial traction, the statement was countered by commentators across the political spectrum who strongly attested to the independence and impartiality of the agency and noted that there was no basis for the assertion of manipulation.

As noted above, OMB issues standards and guidelines that play a critical role in ensuring the integrity, credibility, and independence of the U.S. federal statistical system. Specifically, OMB has directives to ensure that the mission of statistical agencies is adhered to and supported by their parent departments, that statistical releases of information are not subject to manipulation, that appropriate statistical methodologies are used and documented, and that the confidentiality of information is protected (U.S. Office of Management and Budget, 1985, 2006, 2007, 2008, 2014b).

It is also vital that information providers, including businesses and individuals, trust the agency to protect the data they possess from misuse and thereby are willing to provide the requested data (see National Research

Council, 2013b). In the United States, statistical agencies have a superb record of protecting the personal and business data they collect, and there are strong laws protecting information collected under a pledge of confidentiality for exclusively statistical purposes, such as the Confidential Information Protection and Statistical Efficiency Act of 2002 (see Chapter 5). This law and related regulations (see U.S. Office of Management and Budget, 2007) as well as other specific statutes, such as Title 13 for the Census Bureau, are key to respondent trust and ultimately the credibility of the statistical indicators the agencies produce.

All of these laws, directives, principles, and practices are established and widely broadcast to ensure that the relevance and credibility of information produced by statistical agencies is independent from political or other undue external influence (National Research Council, 2013b). Federal statistical agencies strictly follow all of these principles so that their statistical products can provide a solid and objective foundation for policy discussion and decision making.

CONCLUSION 2-1 Federal statistics provide critical information to the country and serve a key role in a democracy.

THE SAMPLE SURVEY PARADIGM

Federal government statistics underwent a revolution between 1930 and 1950 to meet increasing needs for timely information (Duncan and Shelton, 1992). This period saw the development of probability sampling designs and early models to adjust for nonresponse. The 1940 decennial census was the first to employ sampling, with 5 percent of the respondents asked supplemental questions about such topics as the birthplace of their parents, veteran status, and participation in the social security system.[3] The use of sampling in the 1940 census also allowed the Census Bureau to publish preliminary tables about 8 months before the full tabulations were available. The Current Population Survey, the nation's primary measure of unemployment, began using probability sampling for the entire sample in 1943.[4]

Books that were published about probability sampling around 1950 (Cochran, 1953; Deming, 1950; Hansen et al., 1953a, 1953b; Parten, 1950; Yates, 1949) described the advantages of taking a sample instead of conducting a census or using available information. A statistically designed

[3]See http://1940census.archives.gov/questions-asked.asp [November 2016].

[4]The Current Population Survey built on developments in earlier unemployment surveys: the trial census of unemployment in 1933-1934, the Enumerative Check Census in 1937, and the Sample Survey of Unemployment in 1940 (Hansen et al., 1955).

sample allowed estimates to be calculated faster and with less cost because interviewers needed to contact only the sampled households and people and so there were fewer records to be tabulated.[5] One major reason for the adoption of probability sampling was its ability to give error bounds for estimates. Probability sampling uses random selection to draw a sample in which each subset of the population has a known nonzero probability of being selected in the sample. Those probabilities are used to calculate an accurate measure of the precision of the results, and the statistical sampling texts of the 1950s emphasized that the measures of precision in probability samples were based on a rigorous mathematical framework. In contrast, other types of samples—such as convenience samples, which are composed of those people or entities most easily available—also have errors, but those errors are often unobservable and unquantifiable.

The success of probability sampling in providing reliable, trustworthy information led to a proliferation of surveys across the federal government. Many U.S. federal statistical agencies use a sample survey as the default method of producing statistical information. A sample survey can be tailored to particular data needs: it can include the specified questions needed for standardized measurement, and it can include steps to minimize bias from nonresponse and other sources. The agency or its contractor exercises control of the data collection process, often resulting in the collection of high-quality standardized data for a wide variety of characteristics of units and populations of interest. Because of the consistency with which federal statistical agencies use this method, estimates can be compared for different time periods and different locations, which is a crucial feature for such indicators as unemployment and poverty.[6]

Surveys and censuses are currently the principal means of collecting federal statistics. The Census Bureau alone conducts more than 130 economic and demographic surveys every year.[7] In addition, private-sector federal contractors also conduct surveys that are sponsored by federal statistical agencies. As noted above, federal surveys provide vital information on agriculture, the economy, health, crime, transportation, defense, education, energy, housing, social welfare, and virtually every other area in which public policy is set. Each agency often has its own user community and stakeholders.

Although there is tremendous value in the information collected by the federal statistical agencies (see, e.g., U.S. Department of Commerce, 2014),

---

[5]Before the Census Bureau received the very first UNIVAC computer in 1951, tabulations were time-consuming and prone to error. Operators prepared punch cards for data items, which were then processed by Hollerith machines.

[6]Changes in modes and methods do occur, which result in breaks in series, and these are documented to alert all users.

[7]See http://www.census.gov/programs-surveys/are-you-in-a-survey/survey-list.html [January 2017].

there also appears to be redundancy in the collection of information and inefficiency by sole reliance on sample surveys (U.S. Government Accountability Office, 2006). For example, the Current Population Survey, the American Community Survey, the Survey of Income and Program Participation, and many other surveys ask questions about demographic characteristics, income, poverty, and unemployment. For some items and purposes, redundancy is useful: it can add to the reliability of the information and point to sources of variation among different surveys. Yet the redundancy in information collection can lead to additional burden on survey participants, and it can also result in competing national estimates from different sources and different agencies, causing confusion for users. National estimates of health insurance coverage, a key statistic for evaluating the success of the Affordable Care Act, are published by two agencies on the basis of three different surveys: by the National Center for Health Statistics (NCHS) based on the National Health Interview Survey and by the Census Bureau based on both the Annual Survey of Social and Economic Conditions (a supplement to the Current Population Survey (CPS)), and the American Community Survey (ACS). The Census Bureau and NCHS have issued materials to inform and educate users about the appropriate use of each of these estimates and explanations for potential differences among the estimates, but perceptions of burden and confusion about different estimates remains.

In addition to possible redundancy and confusion, there are also questions about some surveys. Although there have been notable improvements in the designs and estimation methods for federal surveys throughout the years, the basic structure of many large surveys has been relatively constant over time. This constancy is due in part to the desire for estimates to be consistently produced from year to year. As noted above, BLS measures unemployment using the CPS and has kept the questionnaire and the statistical methodology for the survey stable to be able to consistently measure changes in unemployment. When the CPS was redesigned in 1994 to update the questionnaire and use laptop computers for the data collection, it had been nearly 30 years since the last redesign in 1967 (Cohany et al., 1994). Because of the importance of this indicator, the change in methodology involved extensive research over several years, including a test running the "old" and "new" surveys in parallel for 18 months to compare results between the new and old procedures to carefully measure the effects of the changes (Cohany et al., 1994; Polivka and Miller, 1995). This example illustrates the careful and transparent research and implementation that is a strength of the federal statistical system. However, it also shows how changes to methods of production of statistics can be difficult and time-consuming, resulting in a system that tends to be conservative and not likely to adapt to new technologies (see Van Tuinen, 2009).

CONCLUSION 2-2 Federal statistical agencies use a highly developed sample survey paradigm for federal statistics.

THREATS TO THE SURVEY PARADIGM

Continued reliance on sample surveys as the principal means of collecting national statistical data is threatened by the increasing difficulty and cost in conducting the surveys, with consequent threats to data quality, and by the increasing demand for more and faster information. Response rates are decreasing for almost all household surveys, adding to the cost of the surveys and, in some cases, raising questions about how well the survey results represent the population. A recent report (National Research Council, 2013a) documented the decreasing response rates and consequences in major federal surveys (see also Brick and Williams, 2013).

Figure 2-1 shows the decrease in response rates for three federal household surveys from 1994 to 2015: the Consumer Expenditure Survey (CES), the National Health Interview Survey (NHIS), and the National Crime Victimization Survey (NCVS). In these three surveys, for at least one interview the data are collected by an interviewer who visits the sample household.

The CES, the oldest of the three, was initiated in 1888; the current form of separate interview and diary components was adopted in 1972. The response rates in 1972 were approximately 94 percent for the interview component and approximately 82 percent for the diary component (Hoff, 1981). These rates decreased to approximately 70 percent for both components in the 2010s despite extensive efforts to maintain them.

The NHIS was launched in 1957. After a household is sampled for the survey and agrees to participate, one adult and one child (if the household has children) are randomly selected to complete the adult and child components. The household response rate in 1963 was 95 percent; in 2015, it was 70 percent. The response rates for the sampled adult and sampled child are lower than that for households because there is additional nonresponse when an individual is asked to participate in the survey: in 2015, the response rates were 63 percent for the sampled children and 55 percent for the sampled adults.

The NCVS, which began in 1972 administers a screening instrument to a household respondent, followed by individual interviews with all people aged 12 and over in the household. Until 1995, the household response rate for the NCVS was consistently above 95 percent, and it has had the smallest declines in response rate of the three surveys. The 2014 household response rate was 84 percent, and that for the interviewed people was 73 percent. The lowest response rates were for nonwhites aged 12 to 24, which is the demographic group with the highest levels of victimization.

Response rate declines have been even more dramatic for telephone

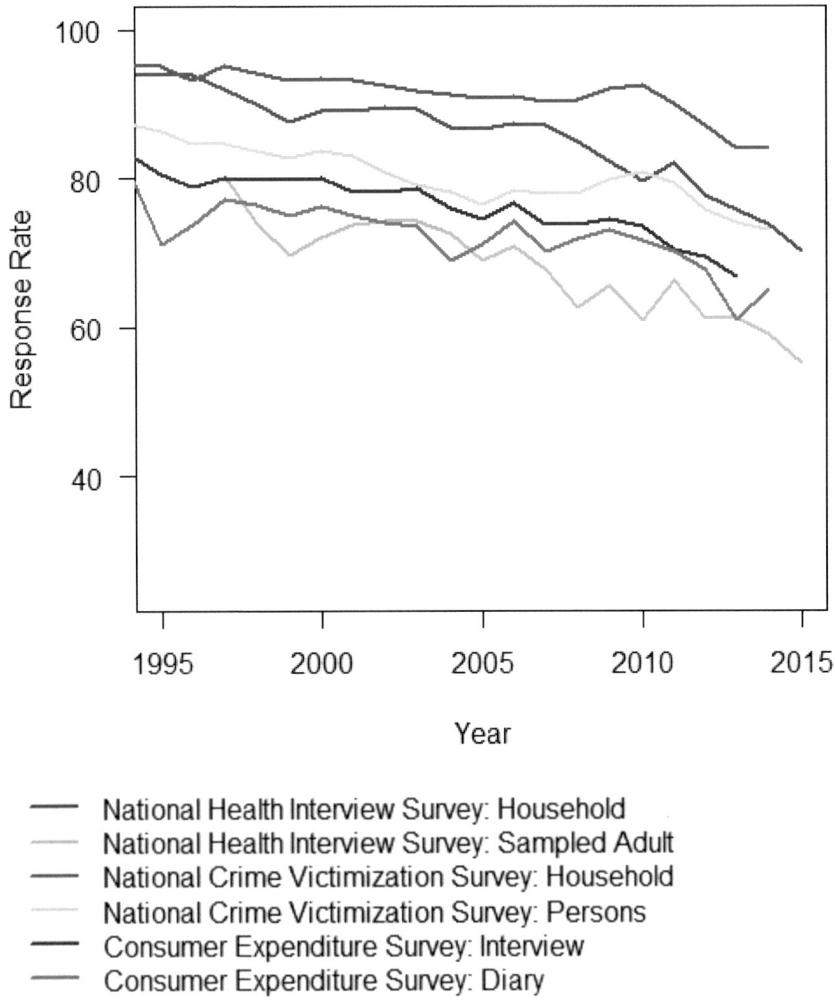

FIGURE 2-1 Response rates for three surveys in which at least one interview is done in person.
SOURCES: Data from National Research Council (2013a, Table 1-1); public-use dataset documentation for the National Health Interview Survey (see ftp://ftp.cdc.gov/pub/Health_Statistics/NCHS/Dataset_Documentation/NHIS/2015/srvydesc.pdf [November 2016]); the National Criminal Victimization Series (see http://www.bjs.gov/index.cfm?ty=dcdetail&iid=245 [November 2016]); and the Bureau of Labor Statistics (see http://www.bls.gov/cex/pumd/2014/csxresponserates.pdf [November 2016]).

surveys. The response rate for the National Immunization Survey (NIS) landline telephone sample declined from 86.8 percent in 1994 to 62.6 percent in 2014 (Centers for Disease Control and Prevention, 2015). And that decline in response rates does not tell the whole story. Before 2011, the NIS called only landline telephones, and the percentage of households *without* a landline increased from 5 percent in 2003 to 32 percent in 2010 (Blumberg and Luke, 2007, 2011). Thus, the NIS 2010 reported response rate of 64 percent refers to responses from landline households alone and ignores the cellular-only and non-telephone households that were not contacted for the survey. Adults in cellular-only households tend to be younger and have different health characteristics than adults in landline households, so any survey that excluded them in 2010 did not represent the full U.S. population (Blumberg and Luke, 2011). In 2011, the NIS implemented a design in which cellular telephones were also called in an attempt to reduce bias resulting from omitting the households that have no landline service. The household response rate for the cellular telephone numbers has been less than 35 percent for every year from 2011 to 2014 (Centers for Disease Control and Prevention, 2015). As a larger proportion of the population becomes cell-only,[8] the overall response rate for NIS is expected to continue to decrease.

Another telephone survey, the National Household Education Survey (NHES), has experienced similar declines in response rates, from 81.0 percent in 1991 to 52.8 percent in 2007 (Grady et al. 2010; Zukerberg, 2010). The NHES discontinued using the telephone as a mode of data collection in 2007 and switched to a mail survey for 2012, which resulted in an increase of the response rate to 73.5 percent (McPhee et al., 2015).

For all three surveys shown in Figure 2-1, the recent large declines in response rates are a disturbing trend because face-to-face surveys often have the highest coverage of the population and response rates, as well as the most control over the data collected (deLeeuw, 2008). The decrease in response rates has led to increased costs as more people must be contacted in order to obtain the required number of respondents to the survey. Decreasing response rates also require other efforts, such as callbacks or mixed mode surveys, which further increase the cost of conducting household surveys.

Similar patterns of decreasing response rates have also been seen in surveys of establishments. In the past 5 years, the Census Bureau has experienced a decline of more than 20 percentage points for the Advance Monthly Sales for Retail and Food Services Survey and a decline of more

[8] As of 2015, 48 percent of adults and 58 percent of children live in households without a landline telephone (Blumberg and Luke, 2016).

than 10 percentage points for the Monthly Retail Trade Survey and the Manufacturers' Shipments, Inventories, and Orders Survey.

In an era of flat or decreasing budgets for many federal statistical agencies (see U.S. Office of Management and Budget, 2015b), efforts to increase response rates have resulted in other declines in the quality of survey work. For the NCVS, for example, increasing survey costs and flat budgets from 1995 to 2006 led to the suspension of interviewer training, reductions in reinterviews for quality control, and cuts in the sample size to compensate for increasing costs. By 2003, the precision of NCVS estimates had decreased so much that the Bureau of Justice Statistics could no longer report reliable estimates of year-to-year changes in victimization, which it had done every year since the beginning of the survey. Thus, the criminal victimization rates for 2003 reported changes in victimization rates between the 2-year averages for 2000-2001 and 2002-2003, instead of 1-year changes between 2002 and 2003 (Catalano, 2004). For the 2006 rates, Rand and Catalano (2007, p. 1) wrote: "The variation in the amount and rate of crime was too extreme to be attributed to actual year-to-year changes."

A review of the programs of the Bureau of Justice Statistics (National Research Council, 2009a, p. 3) found:

> [T]he current NCVS falls short of the vibrant measure of annual change in crime that was envisioned at the survey's outset...[and] as currently configured and funded, the NCVS is not achieving and cannot achieve BJS's legislatively mandated goal to "collect and analyze data that will serve as a continuous and comparable national social indication of the prevalence, incidence, rates, extent, distribution, and attributes of crime."

Following a funding increase, the sample size was increased beginning in October 2010, which gave more precision for year-to-year changes, though at a higher cost.

As response rates continue to decline, there is growing concern about bias from nonresponse, which can occur when the nonrespondents differ systematically from the respondents. As noted above, in the NCVS the demographic group with the lowest response rates is the group with the highest victimization rates. Although adjustments are made to the survey weights to correct for demographic imbalances between the survey respondents and the population and to reduce bias, there is no assurance that the adjustments completely remove nonresponse bias for the key measures of interest about victimization rates. Indeed, the adjustment methods require strong assumptions about the nature of the nonresponse, and as response rates decrease, greater reliance is placed on those assumptions.

Rising rates of item nonresponse, in which a survey participant fails to provide responses to one or more items in the questionnaire, is another threat to the quality of data from surveys. Reports of income can be particularly problematic because respondents are often unwilling or unable to provide this information in detail, which is important for analysis of many federal programs, including Temporary Assistance for Needy Families (TANF), Supplementary Nutrition Assistance Program (SNAP), Supplemental Security Income (SSI), Old Age Survivors and Disability Insurance (OASDI), and Unemployment Insurance (UI). To account for both the level and impact of item nonresponse to a series of specific income questions, one can compute the percentage of total income that is statistically imputed as opposed to reported (see Czajka and Beyler, 2016). Using this approach, Czajka and Denmead (2008) found that more than 50 percent of people in the CPS had some income imputed and that 34.2 percent of the total income in the CPS was imputed. Both the Survey of Income and Program Participation and NHIS had similar amounts of total income imputed, 32.4 percent, while the percent of total income imputed in the ACS was 17.6 percent. The percentage of total income from TANF, SNAP, SSI, OASDI, and UI that has been imputed in the CPS has been increasing over the past 25 years (Meyer et al., 2015). The statistical methods used to impute for missing data rely more heavily on assumptions as nonresponse increases and may introduce bias into the estimates (Citro, 2014). Imputation procedures for earnings in the CPS have consistently underestimated poverty by an average of 1 percentage point due to item nonresponse in earnings (Hoyakem et al., 2014).

INCREASING DEMANDS FOR MORE DETAILED AND MORE TIMELY INFORMATION

The demand for more detailed and timely statistical information has grown steadily in the past two decades (Holt, 2007). The ubiquitous availability of information on the Internet can affect people's perceptions of the timeliness of statistical information. Nowhere was this change in demand more clearly illustrated than in the 2015 reaction of James Comey, director of the Federal Bureau of Investigation (FBI), to his discovering that he could not get an accurate, current, national estimate of the number of citizens shot by the police. After Michael Brown was shot and killed by a police officer in Ferguson, Missouri, Comey wanted to know how many people shot by police were African American, and his staff could not give him this information because it is not collected reliably. He stated that "our data are incomplete and therefore, in the aggregate, unreliable" (Comey, 2015). He later said that it was "embarrassing and ridiculous" that you can "get online and figure out how many tickets were sold to *The Martian*" but "we can't talk about crime in the same way, especially in the high-stakes incidents when your officers have to use force" (quoted in Tran, 2015).

The broad challenge for the federal statistical system is the increased demand for more data more quickly and with more detail for small areas of geography and subpopulations, especially by local governments and businesses. The CPS produces monthly estimates of unemployment and labor force participation, but for many other surveys, estimates are produced annually or less frequently. For the NCVS, the data collection is spread over a year, and the annual estimates of victimization are usually reported about 8 months after the end of data collection. Because most of the people contacted for the NCVS are not victims of crime, the sample sizes for crime victims of specific types tend to be small. Therefore, victimization rates cannot be reliably produced for most states, let alone for metropolitan areas or local jurisdictions.[9] However, it is the local areas where efforts to fight crime take place and where reliable and timely information is needed to take actions.

The ACS is designed explicitly to provide local area information. It is the largest continuous household survey conducted by the federal government and is the replacement for the "long form" from the decennial census. The survey is sent to 3.5 million households every year and includes questions on housing and the demographic characteristics of respondents, as well as such factors as employment, health insurance coverage, and income. Annual estimates are produced for all areas in the country with populations of at least 65,000, while estimates for smaller areas are 5-year rolling averages that are updated annually. Smaller areas include not only smaller cities and towns, but also census tracts within cities and metropolitan areas.

Because the ACS provides comparable data for all areas, its results are widely used by federal agencies for allocating federal program funding, by local governments for local planning, and by private companies for making businesses decisions. The ACS provides much more timely data than was previously available from the decennial census long form (which was available only every 10 years), and these estimates are updated every year. However, the 5-year-period estimates may mask underlying economic and social changes over shorter time periods and make them less useful for understanding current circumstances and informing decisions. The ACS small-area estimates are also less precise than those from the census long form, resulting in greater uncertainty about the current status.

Director Comey's reaction to the lack of needed data is illustrative of

[9]In 2012 the NCVS began a program to provide data-based estimates for the 22 largest states and modeled estimates for smaller jurisdictions. Even with this ambitious program, however, the ability of the survey to make estimates for cities and localities—which are responsible for most crime control policy—is quite limited.

another challenge to the federal statistical system resulting from increased demand—competition from private-sector sources. With the traditional data on police shooting deemed inadequate and not timely enough, the *Guardian* and the *Washington Post* used open-source data to produce estimates of citizen shootings by the police.[10] These ad hoc data collections do not have the stability and transparency of sample surveys, but in the absence of timely "official" estimates from the federal statistical system, alternative estimates of unknown quality and reliability will be used.

> **CONCLUSION 2-3** The way that statistics are currently produced by federal statistical agencies faces threats from declining participation rates and increasing costs. These threats are exacerbated by expanding demands for more timely and geographically detailed information.

---

[10]The results from the two newspapers differed, according to *Politico* (see http://www.politico.com/story/2015/06/police-involved-killings-statistics-washington-post-guardian-118490 [November 2016]).

3

Using Government Administrative and Other Data for Federal Statistics

There has been increasing attention in recent years to "evidence-based policy making," and government statistics are one source of evidence among several diverse tools that have different uses in that endeavor. These tools include impact evaluations, particularly randomized experiments; quasi-experimental evidence from administrative data; observational studies using administrative data, survey data, or linked survey and administrative data; implementation studies; and performance measures (U.S. Office of Management and Budget, 2016). This chapter focuses primarily on administrative records, which the Office of Management and Budget (OMB) defines as data collected by government entities for program administration, regulatory, or law enforcement purpose. They include such records as employment and earnings information on state unemployment insurance records, information reported on federal tax forms, Social Security earnings and benefits, medical conditions and payments made for services from Medicare and Medicaid records, and food assistance program benefits (U.S. Office of Management and Budget, 2014a). In addition to government administrative data, businesses create and keep similar records of transactions and interactions with customers, as well as fulfilling record-keeping requirements for federal, state, and local governments; these data are the subject of Chapter 4.

OMB and the federal statistical agencies have engaged in a number of efforts in recent years to facilitate greater use of administrative records for statistical purposes, with the goal of improving federal statistics and facilitating program evaluation. Statistical purposes are defined as "the description, estimation, or analysis of the characteristics of groups, without identifying the individuals or organizations that comprise such groups" (see P.L. 107-347 §502(9)(A)).

Statistical agencies have worked together to identify and document important case studies that demonstrate the utility of administrative data for statistical purposes and have documented difficulties in being able to access and use administrative data (see Prell et al., 2009). To address those difficulties, OMB issued a memo to all federal agencies that specifically encouraged the use of administrative data for statistical purposes and discussed the legal, policy, and operational issues with using administrative data (U.S. Office of Management and Budget, 2014a). In this memo, OMB encouraged collaboration between program and statistical offices, strong data stewardship policies and practices for the use of administrative data, documentation of quality control measures and key attributes for administrative datasets, and clear designation of responsibilities and practices in interagency agreements.

Several initiatives to improve evidence-based policy making have emphasized the importance of reusing existing government administrative data. These initiatives have included proposals to provide greater access to specific administrative datasets, such as the National Directory of New Hires, as well as to expand infrastructure at the Census Bureau so that it can acquire and process more administrative datasets, expand and improve the process for linking data, and provide access to datasets at the Federal Statistical Research Data Centers (discussed in Chapter 5).

As noted in Chapter 1, a 2016 law established the Evidence-Based Policymaking Commission, whose charge includes the statement that it will (PL-114-140 § 4(a)(1)):

> conduct a comprehensive study of the data inventory, data infrastructure, database security, and statistical protocols related to Federal policymaking and the agencies responsible for maintaining that data to—
>
> determine the optimal arrangement for which administrative data on Federal programs and tax expenditures, survey data, and related statistical data series may be integrated and made available to facilitate program evaluation, continuous improvement, policy-relevant research, and cost-benefit analyses by qualified researchers and institutions while weighing how integration might lead to the intentional or unintentional access, breach, or release of personally-identifiable information or records.

In the rest of this chapter we discuss the benefits and challenges of using government administrative data for federal statistics and describe the use of administrative data in some other countries. We discuss issues of access and other challenges for using data from federal and state and local government programs. We briefly discuss other government data sources, such as information from sensors. We conclude with a brief review of statistical methods for combing survey and administrative data.

BENEFITS

Using program administrative data for statistical purposes provides a number of potential advantages to statistical agencies. Several previous studies have recommended that statistical agencies make greater use of administrative data to evaluate and enhance existing census and survey programs generally (e.g., National Research Council, 2013b) and for specific programs, such as the decennial census (National Research Council, 2004, 2010b, 2011), the American Community Survey (National Academies of Sciences, Engineering, and Medicine, 2016; National Research Council, 2008, 2015), the Survey of Income and Program Participation (National Research Council, 2009b), and science and technology indicators (National Research Council 2010a, 2014a). Because administrative data already exist, the agencies do not incur additional costs for data collection, so they also do not impose an additional burden on respondents. These records typically contain the full population of participants in the program, so the sizes of the datasets are often much larger than those of a statistical survey. For example, the National Health Interview Survey (NHIS) may have only a handful of respondents with a rare combination of medical conditions, but an insurer's electronic health records system is likely to have a much larger number of such people. Furthermore, it may be desirable to combine administrative data with survey data to increase the precision of survey estimates with little additional cost. In other cases, it may be possible to combine different administrative datasets to replace existing surveys or to reduce reliance on survey data. The administrative data are often longitudinal, which enables tracking individuals or businesses over time (see National Research Council, 2009b; Zolas et al., 2015). There are many different kinds of administrative data, which present different challenges for access and use. Some operational data have strictures for the purpose of keeping them secure and accurate for operational uses, so that accessing and using them for statistical purposes can be more difficult than for other administrative records: see Box 3-1.

Administrative data can be used in various ways for statistical purposes, such as:

1. As a survey frame or list of entities, such as businesses or addresses of households. Administrative records may provide a complete frame or a source to supplement an existing frame. A sample can then be drawn from the list to survey.

2. As a replacement for survey data collection if the administrative records include all the information needed.
3. For editing and imputation of survey responses or missing responses. Property tax records could be used to impute information about a dwelling unit that was not reported by the survey respondent.
4. For direct tabulation of administrative records information, such as the number of beneficiaries of a program or the average benefit.
5. As a source of auxiliary information for use in statistical models to improve estimates from surveys.
6. To provide information that could be used to compensate for survey nonresponse.
7. For survey evaluation, such as comparing the total number of program beneficiaries from the program records with the total number based on a survey estimate.
8. For help conducting surveys or census.

BOX 3-1
Using Operational Data for Statistical Purposes

Many of the most interesting data in the area of crime and justice are operational data that are used to conduct routine business in the system. Criminal history data are accessed constantly for background checks and security checks. Strictures for use of these data are designed to keep them secure and accurate for these operational uses. Criminal history data that is over 3 years old is considered to be inaccurate for operational use and must be destroyed. Similarly, expungement of criminal records is seen as a way of reducing the negative effects of criminal records and especially juvenile records.

However, the cultures and demands of operational uses and research and statistical uses are at odds: what is necessary for operations does not apply in the context of statistical uses because statistical data cannot be used to affect the rights, responsibilities, or privileges of individuals. Yet the existing strictures severely limit important research and statistical uses of these criminal justice data. Requiring that data be destroyed after 3 years severely limits the kinds of research questions that can be pursued. Without specific legislation, some valuable operational data will not be available for important research and statistical uses. The Statistics of Income Division of the Internal Revenue Service is able to address this issue by transforming tax data into "statistical records derived from tax returns" that are able to be maintained in perpetuity.

USE OF ADMINISTRATIVE RECORDS BY OTHER NATIONAL STATISTICAL OFFICES

A former director of the Census Bureau noted a key difference between the U.S. federal statistical system and that of many other countries (Prewitt, 2010, pp. 11-12):

> If you ask leaders of the national statistical programs in Europe what proportion of the information they collect comes from administrative data and what proportion comes from survey data, the general answer would be 80 percent from the former and 20 percent from the latter. Ask the same question in the United States and this ratio is reversed.

Indeed, Denmark, Finland, Iceland, Norway, and Sweden use administrative data as the foundation of their whole statistical systems, known as register-based statistics. They are able to do so because of the availability of a number of administrative data sources covering a number of important populations, and a set of conditions that facilitate the extensive use of administrative data. Those conditions include a firm legal basis enabling access to these sources and the requirement to use a unified identification system across sources. These "base registers" contain vital data on people, companies, and addresses/locations. Combining them provides a check on the objects defined by the sources and improves the quality of the whole system.

To these base registers other data sources are linked, usually by the unique identifiers for the various people or entities included. As a result, the majority of the statistics produced by the Nordic countries are largely based on administrative data. Sample survey data are mainly used to provide the vital information not available in administrative sources.

Other European countries rely less than the Nordic countries on administrative records, but still far more than does the United States. Statistics Netherlands has unique identifiers for people, which enable the production of many administrative data-based statistics for people similar to the Nordic approach, including a so-called virtual census (Schulte Nordholt, 2014). However, unique identifiers are not available for companies, but private-sector firms provide their administrative data directly to Statistics Netherlands to produce many economic statistics. This approach requires an elaborate business register in which both administrative and statistical units are related to all business units observed in the real world, which takes considerable effort (Beuken and Vlag, 2010; U.N. Economic Commission for Europe, 2011).

In Canada, the administrative tax data collected by the Canadian Revenue Agency (CRA) has become increasingly important for statistics. It initially formed an important input source for a number of frames, such as the address registers, while it later became important as a (partial) replacement for survey data (Trépanier et al., 2013). For instance, estimates of the total number of employees and gross monthly payroll are based on variables collected by CRA on payroll deduction accounts forms. In a similar way, the CRA provides income tax data that replaces survey data for very small companies and for revenue questions on income for people.

ACCESS TO AND USE OF ADMINISTRATIVE DATA BY FEDERAL STATISTICAL AGENCIES

Federal statistical agencies routinely use administrative data in a wide variety of ways to enhance their survey programs. Those uses include to assist in the construction of sampling frames, to improve the efficiency of the sample design, to impute for missing survey responses, and to weight to known population totals.

The Census Bureau uses federal tax information from the Internal Revenue Service (IRS) to create the Census Business Register, which is the frame for the economic census and most of the Census Bureau's surveys of businesses. The Bureau of Labor Statistics (BLS) creates the frame for all its surveys of business establishments from a different administrative source, the Quarterly Census of Employment and Wages, which BLS obtains from the states, on the basis of the state unemployment insurance program. Many federal agencies and their contractors use an address frame based on information provided by the U.S. Postal Service for surveys of individuals and households.

Although there have been linkages done between surveys and administrative records, such as tax records, for many decades (e.g., Kliss and Scheuren, 1978), federal statistical agencies have more recently been extending these efforts and exploring ways to combine administrative record data with sample survey data and integrate them as part of their regular statistical estimation. In some situations, administrative records could replace surveys; in others, they could be used in combination with surveys to provide more timely, accurate, and detailed information at lower cost. Although administrative data can often be obtained with much less expense than survey data, they are subject to many of the problems that prompted the use of surveys in the first place: the systems of records may not necessarily include all of the population (and it may be unknown which parts are left out); they may be measuring something other than the issue of direct interest; and they may not be collected consistently over time. Thus, it is important to be able to evaluate the quality of these data sources in order to make use of them in the federal statistical system (which we discuss below).

Several federal agencies are already using or planning to use administrative data to improve statistical estimates:

- The Bureau of Justice Statistics, in cooperation with the FBI, is implementing an expanded sample of detailed information from law enforcement agencies for the National Incidence Based Reporting System (NIBRS). Although these administrative data will include only crimes that are reported to the police (so that victimization surveys, such as the National Crime Victimization Survey (NCVS) will still be needed to estimate unreported crime), they will provide a level of geographic detail on reported crimes that is impossible to obtain from the NCVS because of the limited sample size. The annual NCVS dataset typically contains fewer than 800 people who have been the victims of a serious violent crime, while NIBRS includes data on hundreds of thousands of victims of serious violence annually. These two data sources will be used jointly to provide a much more complete picture of crime problems, and the large sample size will make analyses possible that cannot be done in the NCVS.[1]
- The National Food Acquisition and Purchase Survey of the U.S. Department of Agriculture (USDA) uses administrative records from the Supplemental Nutrition Assistance Program (SNAP) to stratify its sample in order to ensure a sufficient representation of SNAP participants in the sample. These same records are used to provide program participation data for corresponding sample households, thereby freeing interviewing time in households for the collection of other data.[2]
- The American Housing Survey (AHS), which is sponsored by the Department of Housing and Urban Development (HUD) (with data collection by the Census Bureau), asks respondents about type of housing, ownership status, mortgage payments, current market value of housing unit, annual real estate tax payments, eligibility for assisted housing, remodeling and repair frequency, and other characteristics of the housing unit, neighborhood, and occupants. A HUD pilot program with the Census Bureau matches the survey respondents with local tax assessment data to research and evaluate the usefulness of the information for the AHS. Because

---

[1]For more information, see http://sites.nationalacademies.org/cs/groups/dbassesite/documents/webpage/dbasse_173129.pdf [December 2016].

[2]For more information, see http://sites.nationalacademies.org/cs/groups/dbassesite/documents/webpage/dbasse_171503.pdf [December 2016] and http://sites.nationalacademies.org/cs/groups/dbassesite/documents/webpage/dbasse_173126.pdf [December 2016].

the assessment data are collected for other purposes, they do not have all of the information of interest for the survey, and different types of information are available in different jurisdictions. There is no clear correspondence between the type of structure information in the survey and that in the assessment data, but the property tax amount is widely available—and often more accurate—in the assessment data than from the survey respondents. Tax assessment data have the potential for replacing some survey items, directly or through statistical models, and for providing data that could be substituted for missing information due to nonresponse.[3]

- The Energy Information Administration (EIA) has been able to produce new statistics on the transportation of crude oil by rail (beginning in 2015) by using administrative data obtained from the U.S. Surface Transportation Board and from Canada's National Energy Board.[4]
- The National Center for Health Statistics (NCHS) replaced two surveys (the National Nursing Home Survey and the National Home and Hospice Care Survey) beginning in 2012 with administrative data from the Centers for Medicare & Medicaid Services on the nursing home, home health, and hospice sectors. NCHS is able to use these data to provide more frequent and more geographically detailed publications of the characteristics of these providers and service users than were possible with the previous sample surveys. Surveys are conducted for other sectors of long-term care, including adult day care and assisted living, where there are no comprehensive nationally representative administrative data.[5]

CONCLUSION 3-1 Administrative records have demonstrated potential to enhance the quality, scope, and cost-efficiency of statistical products.

CONCLUSION 3-2 The use of administrative data can reduce the burden on survey respondents by supplementing or replacing survey items or entire surveys.

Currently, a major barrier to the greater use of administrative records is obtaining access to those records. For example, to create the Longitudinal

[3]For more information, see http://sites.nationalacademies.org/cs/groups/dbassesite/documents/webpage/dbasse_171490.pdf [December 2016].

[4]For more information, see http://www.eia.gov/pressroom/releases/press418.cfm [December 2016].

[5]For more information, see https://www.cdc.gov/nchs/nsltcp/index.htm [December 2016].

Employer Household Dynamics (LEHD) program, which combines administrative data on business establishments and workers with household and business survey data, the Census Bureau had to negotiate separately with every state to obtain its administrative data, through a separate memorandum of understanding (MOU) for each state. The process initially took more than 10 years and requires annual renewals (Abowd et al., 2004). The terms of these MOUs permit the Census Bureau to use the data only for LEHD, not for any other Census Bureau programs.[6] To be able to use these data to improve operations of the decennial census, for example, the Census Bureau would either have to obtain access to the National Directory of New Hires (NDNH) or renegotiate every state MOU. The complexity of getting administrative record data from multiple agencies within states can become even more problematic if an agency or researcher is seeking to build a data warehouse that does not have a specific analysis plan but, rather, wants to provide the linkage and curation of data that others can use. Although this approach would make these data more useful and potentially accessible to various user groups, some state laws, regulations, and common practice do not allow exchanges that do not have a specific research use.

Although OMB requires agencies to look for other sources for the desired information before conducting a survey (see 5 C.F.R. 1320), the statistical agency may not be able to acquire that information from another government agency for a number of reasons. Most often, the reason is that the statistical agency is not authorized by law or regulation to have access to the program agency's data. If so, the statistical agency has little or no recourse since the relevant guidance (U.S. Office of Management and Budget, 2014a) does not compel any program agency to provide the data: it only encourages agencies to work together. For example, the NDNH, which consists of person-level wage records compiled from all 50 states and the District of Columbia (based on quarterly unemployment insurance records), is not accessible to any federal statistical agency because the authorizing legislation for it specifies the agencies and the permitted uses of these data.[7]

---

[6]The narrow interpretation of the Census Bureau's usage in all of the MOUs arose from political influences on state labor market information offices regarding the consequences of state-by-state comparisons on the current state government. LEHD senior scientists led an effort to make changes to IRS Code 6102j to enable a national job frame based on unemployment insurance records, which led directly to the use of 51 state MOUs (in addition to MOUs with the Social Security Administration and the Office of Personnel Management). This use had previously been denied by the 1999-2000 IRS "safeguard review" of federal W-2 data.

[7]The NDNH is compiled by the Office of Child Support Enforcement in the Department of Health and Human Services and used for enforcement purposes, as well as for specific program integrity, implementation, and research purposes (U.S. Office of Management and Budget, 2016).

BLS and the Census Bureau are actively seeking access to this dataset[8] so they are not forced to negotiate separate MOUs with each state to get this information.[9]

The Census Bureau is unique among the federal statistical agencies in that its enabling legislation authorizes it to obtain administrative data from any federal agency and requires it to try to obtain data from other agencies whenever possible (13 USC § 6). However, the statute does not similarly require the program agencies to provide their data to the Census Bureau. In other words, although the Census Bureau is required to ask other agencies for data, they are not required to provide it.[10] The result is that the Census Bureau has been unable to obtain useful administrative data that would simultaneously enhance the quality of various statistical products and reduce burden on the public. In the case of the decennial census, the Census Bureau has provided evidence in its budget submission to Congress that access to administrative datasets, such as the NDNH, is a key element to a cost-effective census that could potentially save billions of dollars in conducting the 2020 census (U.S. Census Bureau, 2015).

One oft-cited advantage of a decentralized statistical system is that statistical agencies are closer to stakeholders and policy makers in their departments than they would be in a single, centralized agency. Although being housed in the same department as program agencies has sometimes made it easier for the statistical agency to obtain administrative records from those program agencies, there are frequently barriers even within departments. For example, the Economic Research Service (ERS) in USDA tried for many years to get SNAP program data from the Food and Nutrition Service (FNS), which is also in USDA. However, FNS does not possess all the program data, which are held by the states. Furthermore, FNS had interpreted the statute authorizing the SNAP program as prohibiting statistical use of program information. Consequently, ERS has not been able to obtain these data from FNS. As a result, the Census Bureau is in the process of negotiating MOUs with each state to obtain access to these data for its statistical use as well as statistical uses by ERS. For another example, the Bureau of Justice Statistics (BJS) is able to obtain access to administrative data from

[8]The President's budget for fiscal 2017 proposes to expand access to the NDNH to specified federal statistical agencies, units, and evaluation offices or their designees for statistical, research, evaluation, and performance measurement purposes associated with assessing positive labor market outcomes.

[9]BLS does have agreements with states for establishment-level data for the Quarterly Census of Employees and Wages (QCEW), but it does not currently get any individual employee-level data.

[10]In contrast, the statistical laws in Canada, Australia, and the Nordic countries require other ministries to provide administrative data to the national statistical office for statistical uses (Statistics Canada, 2016a; U.N. Economic Commission for Europe, 2011).

the Administrative Office of United States Attorneys in the Department of Justice (DOJ) for exclusively statistical purposes, but it is not clear that DOJ would allow access to the data by other statistical agencies.

Another example can be seen in the FBI's Federal Incident-Based Reporting System. The Uniform Federal Crime Reporting Act of 1988 requires that all federal law enforcement agencies (including the Department of Defense) report to the FBI incident-level data on crimes known to these agencies. However, no federal agencies currently share their data with the FBI. Thus, it appears that it is not sufficient to simply require agencies to provide their data.

Even when statistical agencies do have access to administrative data for statistical purposes, those statistical uses can be constrained. The Census Bureau is able to access federal tax information from the IRS for a specified set of purposes (Internal Revenue Code 6103(j)). As noted above, the Census Bureau uses those data to create the Census Business Register; however, BLS does not currently have access to those data and so has to base its frame on a different source. Because BLS and the Census Bureau both conduct different surveys of businesses using different frames, there have been long-standing issues in comparing and reconciling the different statistics that describe the economy from the two agencies (National Research Council, 2007). The Bureau of Economic Analysis (BEA) has acknowledged the differences and cannot resolve them. Being able to use the same business list and synchronize the existing lists would both reduce the burden on businesses and improve the quality of economic statistics, and it is likely that it would also result in cost savings (National Research Council, 2007). The situation is particularly frustrating since BLS and the Census Bureau have had explicit legal authority to allow them to share business information for statistical purposes since 2002 (PL 107-347 Title V, Subtitle B). The required change to the IRS legislation that would permit BLS to have access to limited business tax information has not been passed, despite numerous efforts.[11]

CONCLUSION 3-3 There is currently no agency charged directly by statute with facilitating coordination of access to and use of multiple data sources among federal agencies for the benefit of the entire federal statistical system.

[11]The Obama administration pushed for this legislative authority (see, e.g., U.S. Department of the Treasury, 2014; U.S. Office of Management and Budget, 2016), but despite support from previous administrations and broad support from the statistical and research community no action has been taken for this limited data sharing of business tax information for exclusively statistical purposes by Census, BEA, and BLS (see http://www.copafs.org/UserFiles/file/FederalBusinessRegistryLetterSenatewithAttach.pdf [December 2016]).

CONCLUSION 3-4 Legal and administrative barriers limit the statistical use of administrative datasets by federal statistical agencies.

We discuss approaches for addressing these barriers of access to administrative datasets in Chapter 6.

STATE AND LOCAL GOVERNMENT ADMINISTRATIVE DATA FOR FEDERAL STATISTICS

In the United States, much administrative data relevant to national problems and issues are collected and owned by states and localities. When these data are used for statistical purposes, federal statistical agencies aggregate them to produce national estimates. Some data are available for and used in federal statistics; other data that could be valuable are not.

One example of available data is information on the U.S. prison population, the vast majority of whom serve their sentences in state correctional facilities, and state departments of correction collect data on these populations. This information includes basic descriptive data on the people admitted to and released from the institutions: conviction offense, date of admission; date of release; the person's age, race, and ethnicity; the offender's criminal justice status at admission; and the county of conviction. These data are sent to BJS to provide basic descriptive statistics on the state of correctional populations nationally.[12] These data are used to design federal programs that inform prison construction as well as programs for the reintegration into civilian life for prisoners who have served their sentences.

Educational statistics have a similar local-state-federal organization. Since 2005, for example, the National Center for Educational Statistics in the U.S. Department of Education has been building the Statewide Longitudinal Data System (SLDS), which is an integrated system of data on student achievement and performance at the state level. Grants are made to state departments of education to design, develop, and implement a system of administrative records that tracks student achievement in the state. These grants are used to obtain the personnel, hardware, software, and technical assistance for states to collect and share information on students over time. In return, states are expected to share these data with the Department of Education and make them available for research. After a decade of effort, 47 states are participating in SLDS.[13]

In the field of criminal justice, the National Instant Criminal Background Check System (NICS) uses data from state and local police, courts,

[12]See https://www.bjs.gov/index.cfm?ty=dcdetail&iid=268 [December 2016].

[13]See https://nces.ed.gov/programs/slds/about_SLDS.asp [December 2016].

and correctional institutions to construct a criminal history for anyone booked for an offense in any state. Originally, this information was sent to the FBI's National Crime Information Center, but since the Brady Act in 1993 the states have maintained this information in their own repositories. These repositories are linked by the Interstate Identification Index, which serves as a pointer system to state repositories with criminal history records for a particular individual.

The Brady Act prohibited convicted felons and other offenders from being able to purchase a firearm, and NICS is the system by which required background checks are performed. The use of background checks with the NICS data has grown to include not only firearms purchases, but also checks for applicants for many occupations and licensing. BJS has used the NICS data to estimate the rate of recidivism of state prison inmates, a crucial indicator of the success of state correctional programs. Recidivism data are also essential for the development of risk assessment tools that can help to reduce the use of confinement and incarceration. The NICS data are currently being linked with information from the Survey of Inmates of State Correctional Facilities in an effort to understand how the experience of inmates during their imprisonment affects their likelihood of subsequent offending.

As noted above, the LEHD program combines administrative data from the states with Census Bureau census and survey data through the Local Employment Dynamics Partnership. Under this partnership, states agree to share unemployment insurance earnings data and QCEW data with the Census Bureau, and the survey and administrative data are used to create statistics on employment, earnings, and job flows at detailed levels of geography and industry and for different demographic groups. In addition, the LEHD program uses these data to create partially synthetic data on workers' residential patterns. Currently 49 states, the District of Columbia, Puerto Rico, and the U.S. Virgin Islands have joined the LED Partnership.

Federal statistical agencies routinely request administrative record data from states and localities, but state and local agencies are usually under no legal obligation to provide them given the country's constitutional guarantees of the independence of these subnational units of government. For example, the Uniform Crime Reports (UCR) uses information from local police departments to estimate the national rate of crimes known to the police; however, there is no federal law requiring states and localities to report these data to the FBI, and participation is not universal. In contrast, states can require localities to report UCR crimes to the state, and more than 25 states do have such laws (Bureau of Justice Statistics, 2003). The state laws undoubtedly help the UCR achieve a very high response rate for offenses known to the local police departments.

Without national legal requirements to share state and local administrative data, various incentives have been used to encourage data sharing by states and localities, with varying degrees of success. Principal among these incentives is making the receipt of federal program funds dependent on sharing state and local administrative record data with the funding agencies. However, using federal program funding as an incentive to encourage data sharing is not without its problems. At the least, "coerced compliance" may be minimal compliance rather than a robust partnership.

One example is in education. The reporting requirements for postsecondary education institutions under the "Clery Act" (20 U.S.C. § 1092(f))—which requires colleges and universities who receive federal funding to share information about crime on and around campuses—are extensive, and the penalties for noncompliance are severe. Fines for noncompliance have been in the hundreds of thousands of dollars, and there is an extensive apparatus for monitoring compliance. Nonetheless, the data on sexual violence reported by many institutions in response to the act's requirements is of questionable quality (Becker, 2015; California State Auditor, 2015).

It is important to consider how a better incentive structure or federal-state partnerships could operate to optimize sharing of state and local administrative data on a larger scale. Without an improved incentive structure, sharing of state and local administrative data is not likely to be effective. It is equally important to recognize that in many cases it is not that the states and localities are unwilling to share data—they are unable to do so. They do not have the information technology infrastructure to comply with data-sharing requests or to change the collection, coding, or format of data to comply with national reporting standards. They do not have information technology and statistical staff who could comply with the data requests while performing their regular activities. When the FBI and BJS attempted to implement NIBRS for the first time in the 1990s, the greatest challenge to participation for states and localities was lack of resources to replace outdated management information systems and to make required changes in the more modern systems (Roberts, 1997).

An increasingly popular strategy for encouraging the sharing of administrative record data from states and localities is a grant program for building the infrastructure required for sharing, standardizing, protecting, and disseminating administrative record data. The effectiveness of this strategy is likely to depend on the audience and the application; the audience would have to want the information and analysis, and its application would need to be important. The SLDS, described above, follows this strategy in education. State departments of education are awarded grants to integrate state administrative records on student achievement and ultimately share these data with the federal Department of Education. These grants pay for the hardware, software, and technical assistance necessary to link data on schools' and students' achievement over time.[14]

BJS and the FBI are collaborating in building the National Crime Statistics Exchange (NCS-X), which is a sample-based collection of incident-level administrative record data on crimes known to the police; it is designed to replace the UCR summary reporting system (see above), which has not been substantially updated since 1929. NCS-X is providing grants for hardware, software, and technical assistance for jurisdictions to convert their records management system for incident reporting from that needed for the UCR. It will be important to see whether the sharing lasts beyond the life of the grant program.[15]

In contrast to programs like NCS-X and SLDS, which use the building of state and local infrastructure to encourage the exchange of administrative record data with the federal government, there are other programs that use the quid pro quo of providing enhanced data back to the states and localities. LEHD uses this approach to cement the exchange of administrative record data with the states. The unemployment insurance data from the states are linked with other states and census data to provide a picture of local labor markets that is much more complete than that which the states could do on their own, such as being able to track graduates of state colleges and universities when they move to different states to find jobs.

A similar incentive structure is offered by the Center for the Administrative Records Research and Analysis (CARRA) at the Census Bureau. CARRA can perform data linkages "behind the wall" at the Census Bureau to permit expanded statistical uses of existing administrative and survey datasets. BJS provides release records from state departments of correction, under the National Corrections Reporting Program, so that they can be matched to Social Security data and ACS data for statistical purposes. However, freedom of access to these data without new permission from the original contributors varies by dataset.

CARRA not only has the advantage of having a broad range of data that could be used to enhance the data contributed by states and localities,

---

[14]For more information, see https://nces.ed.gov/programs/slds/grant_information.asp [December 2016].

[15]Plans are under way to sustain the exchange after the original infrastructure is built, through mutually beneficial arrangements. For example, tools would be available, through "cloud" storage and computing, that would allow for analyses of crime rates and police response across jurisdictions for participating police departments. In addition, there are also mandatory reporting laws at the state level for the Summary UCR system that can be adapted to include NIBRS reporting and thereby perpetuate the information exchange. At this stage in the development of NCS-X, the emphasis is on making grants available to localities for modifying their management information systems so that they can provide incident-level data to the NCS-X program.

it also has the ability to link these data at the microlevel in a secure environment. Such linking simplifies the confidentiality problems attendant to linkage in other environments that require data to be transmitted from the owner to the user before linkage can take place. CARRA uses a protected identification key (PIK) that can be used to link any dataset in its holdings. When datasets are received, each person in the dataset is associated with a PIK, which is an encrypted identifier. Having a PIK permits accurate linkage to a wide array of Census Bureau data and other data collections.[16] (In other environments, identifiers would need to be produced uniquely for each pair of matched datasets.)

Although there are decided advantages in enhancing the value of state and local data and using this quid pro quo to encourage data exchange, it may not work in policy domains in which state research traditions are not robust. Traditions of science-, evaluation-, and evidence-based policy are deeply entrenched in medicine and education, for example, but much less so in law enforcement and the judiciary. Entities that own the data are much more likely to exchange their administrative records in return for research and evaluation or enhanced data in science-based domains than other domains. And enhanced data may not be much of an incentive in domains in which empirical evidence does not have a strong history. However, there are good models for how states and local governments integrate multiple sources of their administrative data, both to improve program operations and facilitate important policy research (see the example in Box 3-2).

Establishing a stable exchange of administrative records with states and localities is of paramount importance for using these data sources for federal statistics. But the problems of establishing and enforcing uniform national standards for reporting, of protecting the confidentiality of the data, and of developing standards for fitness of use all need to be addressed. As individual programs learn what works in these efforts at building systems for sharing some administrative records between levels of government, it could be helpful for federal statistical agencies to have a concerted effort to share the lessons learned. Many of these efforts are proceeding in isolation and confront similar problems that may be resolved more easily with a common solution. Even at this early stage of development, it could be useful to have a forum for federal statistical agencies to share their experiences in exchanging administrative records with states and localities. The panel views incentives for state and local authorities not as a simple solution to the issues of obtaining access to their administrative records for statistical purposes, but as a necessary, though not sufficient, condition to improve federal statistical agency access to those records.

---

[16]For an examination of errors associated with the use of PIKs, see Abowd and Vilhuber (2005).

BOX 3-2
Use of Administrative Records by Local Governments

The Integrated Database on Child and Family Programs (IDB) in Chicago was created to address needs for a data infrastructure to address policy questions. Researchers from Chapin Hall, a research center at the University of Chicago, partnered with state, county, and city agencies to bring many different administrative datasets together, including data on maternal and child health, the infants and children nutritional program, SNAP and food stamps, Medicaid, abuse and neglect reports, child welfare services, incarcerations, arrests, employment, and earnings. These data are linked at the individual level, and often these data are multigenerational, so that children, parents, and often grandparents are linked. Chapin Hall is responsible for the storing, quality, and technical aspects of the datasets to ensure that they can be linked for research purposes.

Researchers at Chapin Hall and other institutions have been able to use the IDB to create peer-reviewed journal articles, as well as formal reports for policy makers. The governor's office of Illinois requested a study to identify families in multiple systems and multiple programs to better understand the costs associated with program utilization. The results of Chapin Hall's analysis showed that 23 percent of extended families that were being served in multiple systems were accounting for 64 percent of service intervention resources and 86 percent of funding resources. Spatial analysis of the data further showed these families tended to live in geographic pockets. These results allowed Illinois to target service delivery resources geographically and individually (Goerge et al., 2010).

CONCLUSION 3-5 State and local governments may respond to incentives from the federal government to provide access to their administrative data by federal statistical agencies for statistical purposes.

CONCLUSION 3-6 Federal statistical agencies could benefit by sharing their experiences exchanging administrative records with states and localities. This could be done through a forum or interagency working group in which they could seek common solutions and identify incentives for states and local governments to provide access to their administrative data.

CHALLENGES IN USING ADMINISTRATIVE DATA FOR FEDERAL STATISTICS

Once a federal statistical agency has gained access to an administrative dataset, it has to evaluate the data for its utility for potential statistical uses. Because administrative data are collected by government entities for program administration, regulatory, or law enforcement purposes rather than statistical purposes, they may in their "raw" form not be suitable for statistical purposes for a variety of reasons. Administrative data can have many limitations, including (1) lack of quality control, (2) missing items or records (i.e., incompleteness), (3) differences in concepts between the program and what the statistical agency needs, (4) lack of timeliness (e.g., there may be long lags in receiving some or all of the data), and (5) processing costs (e.g., staff time and computer systems may be needed to clean and complete the data).[17]

We note in Chapter 2 that federal statistical agencies and the survey research profession more generally have developed a sophisticated framework for examining and evaluating the quality of data from surveys. A similar framework is needed to understand the quality of administrative data. Some other national statistical agencies have created quality frameworks for their administrative data (see, e.g., Daas and Ossen, 2011; U.N. Economic and Social Council, 2014). We will review these frameworks in more detail in our second report.

In addition to those considerations, statistical agencies also need to ensure that the general public appreciates and understands the benefits of using administrative data for federal statistics and that there is broad public approval of this use. Following the *Principles and Practices for a Federal Statistical Agency* (National Research Council, 2013b), discussed in Chapter 2, there should be transparency in the way statistical agencies use administrative data. The Administrative Data Research Network (2015) in the United Kingdom has produced informative videos to clarify how data are handled and people's privacy is protected (we discuss privacy issues in Chapters 5 and 6).

CONCLUSION 3-7 Not enough is yet known about the fitness for use of administrative data in federal statistics. Coverage, missing information, lack of consistency, and continued availability present challenges with their use.

RECOMMENDATION 3-1 Federal statistical agencies should systematically review their statistical portfolios and evaluate the potential benefits and risks of using administrative data. To this end, federal statistical agencies should create collaborative research programs to address the many challenges in using administrative data for federal statistics.

[17]Additional challenges include spatio-temporal-demographic misalignments. For a detailed discussion of these issues, see Lavallée (2000).

OTHER GOVERNMENT DATA SOURCES

We note briefly that there are other kinds of new data, in addition to administrative data, that are held by federal, state, or local governments. Some examples are weather conditions and water quality data from sensors, videos from traffic cameras, and geospatial data (see Table 3-1). As with administrative records, these other data are not created with the primary intention of statistical use, yet they may provide valuable information for official statistics. However, they have even greater challenges than administrative data for such use. These other data vary in their readiness for statistical uses; some data sources are much more organized and structured than others and amenable to statistical analysis. Government weather data, for example, often comes in very structured forms that can easily be incorporated in a database and used for modeling or analysis; in contrast, videos from traffic cameras are much more unstructured in nature.

For example, in New York City, the meters in taxis—which include how long each trip takes, the start and end points for each trip, the time of day the trip was made—can be used in transportation statistics that show how often traffic jams occur, what time of day roads are most traveled, which season has the heaviest traffic, or how many customers travel to and from the airports. These data can be integrated with weather data to examine anomalies in traffic patterns. When comparing annual taxi meter data, it may not seem strange that certain days have a very low number of rides in comparison with other days, such as Christmas. However, some days may have much lower ride rates compared with the same days in other years. For example, late October 2012 had very low ride rates due to Hurricane Sandy (Freire et al., 2016).

Statistical agencies have been exploring some of these data sources. The National Agricultural Statistics Service has been exploring the use of geospatial data, weather data, and other environmental data in models with survey data (Cruze, 2015). We discuss the challenges with these other data sources in Chapter 4.

COMBINING SURVEY AND ADMINISTRATIVE DATA SOURCES

There are many reasons, some of which are noted above, and methods for combining survey and administrative (and other) data. In the list below, we briefly note four of them, based on Citro (2014) and the review of statistical methods for combining data in Lohr and Raghunathan (in press).

1. Link records at the person or entity level across data sources. As noted above, CARRA has developed a system that assigns unique identifiers to people in the decennial census, federal surveys, admin-

TABLE 3-1 Types and Examples of Government Data Sources

Definition and Examples	Structured Data: Censuses and Probability Surveys	Structured Data: Administrative Records	Other Structured Data	Semi-Structured Data	Unstructured Data
Definition	Collecting data from the universe or a sample of that population and estimating their characteristics through the systematic use of statistical methodology	Data collected by government entities for program administration, regulatory, or law enforcement purpose	Data that are highly organized and can easily be placed in a database or spreadsheet. They may still require substantial scrubbing and transformation for modeling and analysis.	Data that have structure, but also permit flexibility in structure so that they cannot be placed in a relational database or spreadsheet. Transformation into structured form requires decisions with regard to the way in which to standardize the observed variety in structure. The scrubbing and transformation for modeling and analysis are usually more difficult than for structured data.	Data, such as in text, images, and videos, that do not have any pre-defined structure. Information of value must first be extracted from such data, after which the extracted information can be placed in a structured table for further processing and analysis

Government Examples	• Decennial Census of Population and Housing
	• Economic Census
	• Agriculture Census
	• Federal statistical surveys
	  ○ ACS
	  ○ CPS
	  ○ NHIS
	  ○ AHS
	  ○ NCVS
		• Federal records
		  ○ Income tax
		  ○ Social Security
		  ○ Unemployment
		  ○ Medical records
		• State records
		  ○ SNAP data
		• Police accident reports
		• County records
		• Other jurisdictions' data
			• Weather sensors
			• Traffic sensors
			• Water quality sensors
				• Web-scraped quantitative data
				• Web logs
					• Traffic videos
					• Satellite images
					• Blogs and comments
					• Input in free text fields

istrative records, and commercial data (Wagner and Layne, 2014), which allows records in different sources to be linked. The linked records of people who are in multiple data sources then have all the measurements taken in the different sources: they may have tax information from one source, health outcomes from a second source, and education information from a third source. An analyst can study relationships among tax, health, and education information—relationships that could not be studied if only a single source was used. Data linkages can be used to fill in missing information from some sources or to correct erroneous information. Even if some data sources contain people not found in the other sources, the combined data file contains more members of the population than any of the original sources considered individually.

2. Use information from administrative records or other sources to improve the design of probability surveys or the accuracy of estimates computed from them. Almost all surveys use information from external sources to develop the sampling frame, to stratify the sample designs, and to improve the precision of survey estimates. As an example, the Current Population Survey (CPS) uses information from the decennial census and other sources to stratify the sample and determine optimal probabilities for selecting units to be in the sample (Bureau of Labor Statistics and U.S. Census Bureau, 2006). These design features allow the CPS to achieve higher precision without increasing its cost. The CPS also adjusts the weights of sampled units so that survey estimates of the total number of people belonging to age, race, ethnicity, and sex groups are forced to equal independent counts of those groups in the population.
3. Combine statistics that are calculated from different probability surveys or from other sources. Probability samples are typically selected from a sampling frame that describes the population to be sampled, but different frames may include different parts of the population of interest so that taking samples from multiple frames may cover more of the population of interest. A sampling frame that consists of landline telephone numbers will not cover people who have only cell phones or people with no telephone service. When estimates from a sample of landline telephone numbers are combined with estimates from an independent sample of cell phone numbers, the survey results can represent both landline and cell phone households. Using two sources for sampling frames, which have very different costs, can also permit an inexpensive source to provide detailed information on part of the population while still having representation of people who are only included in the more expensive source.

4. Use statistical models to combine information from different datasets. Many types of statistical models can be used to combine datasets, ranging from weighted averages of estimates to imputation models and hierarchical Bayesian analyses. Examples of using hierarchical models to combine information across data sets are given by Cruze (2015), Giorgi et al. (2015), Manzi et al. (2011) and Schenker and Raghunathan (2007). Many of these models allow assessment of the variability that may arise because different sources use different methods or question wording to collect data. The Small Area Income and Poverty Estimates program uses statistical models to combine information from the American Community Survey (ACS) with information from administrative records (see Box 3-3).

The statistical models used to combine data sources make strong assumptions about the relationships between variables across the different data sources. For example, the models often assume that relationships that hold for records or areas present in multiple data sources also hold for records or areas present in just one of them. An advantage of the modeling approach, though, is that these assumptions can be stated explicitly and can sometimes be evaluated empirically. We will examine these models, their assumptions, and their potential use to enhance federal statistics in greater detail in our second report.

CONCLUSION 3-8 Combining multiple datasets allows for expansion of the number of attributes on people or entities and thus can improve federal statistics, including the capacity to perform multivariate analysis for policy and evaluation studies.

CONCLUSION 3-9 There are statistical methods and models for combining information from multiple data sources using a variety of techniques.

There are many challenges to be addressed and risks to confront in using administrative and other data sources for federal statistics, and sophisticated statistical methods will not address all of them. As Louis (2016, p. 20) has noted, "Space-age procedures will not rescue stone-age data." Data sources need to be carefully vetted, and further developments are needed in quality frameworks and statistical methods. Indeed, a combination of statistical methods may be needed for making optimal use of multiple data sources. A framework is needed for combining different data sources so as to draw on the strengths of each source while counterbalancing that source's weaknesses. Such a framework needs to include several elements:

BOX 3-3
Small-Area Estimation

The Small Area Income and Poverty Estimates (SAIPE) program at the Census Bureau produces state- and county-level estimates of poverty by combining estimates from the ACS with estimates from other data sources. The county-level estimate of number of people under age 18 in poverty is calculated as a weighted average of the ACS estimate for the county and a predicted value that relies on information from tax returns, state data files for SNAP, population projections, and estimates of poverty from the 2000 census (the most recent census in which poverty was measured).[a]

The statistical model allows estimates to be produced for every county, even for counties for which the ACS provides almost no information about the poverty rate. When the ACS sample size in the county is small, the estimated number of people under age 18 in poverty relies almost entirely on the prediction from the other data sources. This type of modeling, called small-area estimation, is used throughout the federal statistical system. Examples include estimating state- and county-level diabetes prevalence,[b] labor force properties,[c] and health insurance coverage.[d] Small-area estimation methods leverage information from other sources to provide estimates at a level of geographic detail that would not be possible from the survey data alone. The SAIPE program uses administrative data to estimate county-level poverty rates, but any external source of information can be used as a predictor variable in a small-area model, including electronic health records from medical providers, satellite-measured reflectance, sensor data on traffic flow, or detailed information from cell phone providers.

---

[a]See http://www.census.gov/did/www/saipe/data/highlights/files/2015highlights.pdf [December 2016].
[b]See https://www.cdc.gov/diabetes/data/index.html [December 2016].
[c]See http://www.bls.gov/lau [December 2016].
[d]See http://www.census.gov/did/www/sahie/index.html [December 2016].

- Methods for assessing the error qualities of a data source, including aspects of coverage (Who is missing from the data source?), measurement error (Do the data differ from the "truth"? Do the data differ from what the investigator wants to measure?), and nonresponse bias (Do respondents differ from nonrespondents?), as well as sampling error. The assessment needs to include consideration of the stability of the data source over time and any potential for the source to be manipulated by outside interests.
- Assessment of the accuracy of methods used to combine data sources. Data linkage methods can augment the amount of information available by making use of multiple data sources, but errors

can occur if there is insufficient information to allow accurate linkage (see Herzog et al. [2007] for a discussion of the impact of linkage errors). A person's name may be listed as "Michael" in one source and "Micky" in another so that the records belonging to the same person are not linked. Other records may be erroneously linked, such as linking records for "Michael Smith" may actually be for different people. For the CARRA system, Wagner and Layne (2014) found correct matches for more than 90 percent of the records in the 2010 census and more than 70 percent of the records in two commercial files, but match rates for other sources can be much lower. Further evaluation of the CARRA data linkage methodology using PIKs is needed.

- Statistical models used to combine information from different sources often have strong assumptions about the properties of the data sources or the relationships among variables and can produce erroneous estimates if those assumptions are not met. A robust program of assessing the sensitivity of estimates to model assumptions is needed. In addition, models need to be updated and continually improved to describe current relationships among variables.
- The measures of uncertainty about estimates produced by combining data sources need to include the error properties of the data sources and the statistical methods, in addition to the measures of traditional errors based on sampling theory. More research is needed to improve the uncertainty measures for estimates based on combining multiple data sources.

CONCLUSION 3-10 Dealing with multiple data sources is more complex than dealing with a single dataset. A framework is needed to identify the error structure of each source and assess the utility of combining different data sources given their strengths and weaknesses.

In our second report we will discuss in greater detail an error framework for estimates based on combining multiple data sources, as well as the potential implementation of these methods in ongoing production systems of federal statistical agencies. We will also further describe areas for research and development.

4

Using Private-Sector Data for Federal Statistics

Recent years have witnessed an explosion of data from many sources, some of which are referred to as "big data" (e.g., Daas et al., 2015; Manyika et al., 2011). The term refers to the vast amounts of data that are now available in electronic form and are potentially accessible to analysis, including data that previously existed but were not centrally accessible (such as sales data and medical records) and new kinds of data for phenomena that were not previously measured on a consistent basis but now can be, using new kinds of measurements (such as sensors of natural and artificial phenomena—weather and traffic). IBM has estimated that 2.5 exabytes (2.5 million terabytes) of data are produced every day.[1] As a comparison, the U.S. Library of Congress has roughly estimated that its entire printed collection of 26 million volumes totals 208 terabytes.[2] Some of these new data come from digital records of government agencies (e.g., the health care transaction records of the Centers for Medicare & Medicaid Services). But many of them come from private-sector enterprises (e.g., Manyika et al., 2011). Indeed, a whole set of new enterprises are using large digital data resources as the basis of their business models (e.g., Uber, AirBnB, LinkedIn).

For this report's purpose, we consider two kinds of private-sector data: private-sector structured data and private-sector data that have high dimensions, either in the number of observations or records or the number of

---

[1]See https://www-01.ibm.com/software/data/bigdata/what-is-big-data.html [November 2016].

[2]See https://blogs.loc.gov/thesignal/2012/04/a-library-of-congress-worth-of-data-its-all-in-how-you-define-it [November 2016].

attributes of the observations. Examples of high-dimensional data include streaming data production (e.g., utility meters, traffic cameras, and other sensors), Internet behavior documentation (e.g., browser search terms), and social media postings (e.g., data from Twitter, Facebook, LinkedIn). Examples of structured data include consumer information data, such as those from Zillow and Experian and other credit bureau data.

Some of these new data—whether from government or private-sector sources—could be used to create new statistics by themselves; others could be and are being using in conjunction with traditional statistical data. Some are stored in a form that permits useful statistical analysis immediately; others are stored in forms that would require significant processing prior to their statistical use.

In this chapter we first review the different kinds of private-sector data that are available and how the characteristics of these data affect their potential utility and usability for federal statistics. Next we briefly review efforts by national statistical offices around the world to examine and experiment with using these data sources to produce official statistics. We then review current work in the United States to examine and evaluate these new data sources for federal statistics. We conclude the chapter with a discussion of the challenges in using these data for federal statistics, including issues of access and quality.

DIMENSIONS OF NEW DATA SOURCES

We distinguish three dimensions of the new data resources: who owns or controls them (i.e., government agencies (federal, state, local) or private-sector entities), the purpose for which the data were created (e.g., record transactions or output from sensors or to communicate with others through social media), and the form of the data as stored (i.e., structured numeric data, semi-structured data, unstructured text, pixel data). In this chapter we deal primarily with private-sector data. Table 4-1 details these categories of new data resources.

The data sources shown in Table 4-1 vary in their "readiness" for use in federal statistics in terms of the likely time and effort it would take to clean and format them in order to produce usable statistics. As the second column of Table 4-1 indicates, private firms use surveys to assess their customers' satisfaction or conduct broader surveys of a target population for market research or to make estimates of media use (e.g., Nielsen). The weaknesses in the survey paradigm (see Chapter 2) have also become very evident to private survey firms, which have generally lower response rates than do government surveys. In fact, many firms have abandoned the probability survey paradigm for opt-in Internet panels (Baker et al., 2010).

TABLE 4-1 Types and Examples of Private-Sector Data Sources

Definition and Examples	Structured Data from Censuses and Probability Surveys	Structured Data from Administrative Records	Other Structured Data	Semi-Structured Data	Unstructured Data
Definition	Data from a population or a sample of that population used to estimate the population's characteristics through the systematic use of statistical methodology	Data collected by private companies from transactions, process control, or financial or human resource records	Data that are highly organized and can easily be placed in a database or spreadsheet, though they may still require substantial scrubbing and transformation for modeling and analysis	Data that have structure, but also permit flexibility in structure so that they cannot be placed in a relational database or spreadsheet; the scrubbing and transformation for modeling and analysis is usually more difficult than for structured data	Data, such as in text, images, and videos, that do not have any structure so that information of value must first be extracted and then placed in a structured table for further processing and analysis
Private-Sector Examples	• Customer satisfaction surveys • Marketing research surveys • Media use surveys • Academic surveys	• Data produced by businesses ○ Commercial transactions ○ Banking and stock records ○ Credit card records ○ Medical records • University and other nonprofit grant transactions	• E-commerce transactions • Mobile phone location sensors • Global Positioning System sensors • Utility company sensors • Weather, pollution sensors	• Extensible Markup Language (XML) files • Data from computer systems ○ Logs ○ Web logs • Mobile phone content: text messages • E-mail • Internet of things[a] • Sport activity sensors (from watches, etc.)	• Social network data (Facebook, Twitter, Tumblr, etc.) • Internet blogs and comments • Documents • Pictures (Instagram, Flickr, Picasa, etc.) • Videos (YouTube, etc.) • Internet searches • Traffic webcams • Security/surveillance videos/images • Satellite images • Drones • Radar images

[a]The Internet of things refers to electronics embedded in devices and machines that allow them to be connected to the Internet to directly send and receive data.

In addition to government and private-sector data, surveys and censuses are also conducted by academic researchers. The data from these surveys are sometimes combined with administrative records to produce valuable information. For example, the Health and Retirement Study, conducted by the University of Michigan, obtains earnings records from the Social Security Administration and Medicare claims and summary information from the Centers for Medicare & Medicaid Services that are matched to respondents' survey data to produce statistics and analysis about Americans' physical and financial well-being.

As shown in the third column of Table 4-1, private firms also generate their own administrative records, which may be similar in structure to government administrative records. In the private sector, administrative records are often transactions, such as credit card purchase records or payroll documents. Sometimes these private-sector administrative records are used to produce statistics on their own, such as the National Employment Report from Automatic Data Processing, Inc. (ADP), which precedes the Bureau of Labor Statistics (BLS) release of the employment situation each month.[3]

The other three categories for private-sector data sources, shown in the last three columns of Table 4-1, vary in the structure of the data and how difficult they are to clean and transform into usable numeric form to produce statistics. By structured data we mean numeric data, often ordered into rectangular or fixed relational formats. The best structured data for statistical use have metadata attached to them, which document the format and meaning of each variable. However, even with these attributes, structured data generally need to be transformed for analytic purposes. Structured data in the private sector often include highly detailed geospatial data, such as those from mobile phones, traffic sensors, and Global Positioning System (GPS) devices, and these data may be available in real time. Some similar sensor data, including traffic monitoring sensors, may also be created by government agencies (see Table 3-1 in Chapter 3).

Semi-structured data can be best described as data that can be turned into formatted numeric data by being coded and classified into numeric categories based on information available from the unstructured data themselves. Examples of semi-structured data include Extensible Markup Language (XML) files, e-mails, documents, mobile data content, and log data from computer systems.

Unstructured data include digital videos, digitized pictures, and digital sound recordings, as well as digitized text. Some common forms of private-

[3]ADP works in collaboration with Moody's Analytics in using ADP's large payroll dataset to predict private-sector employment prior to the BLS release. ADP processes the payrolls of about half a million private establishments in the United States, which employ nearly 20 percent of private-sector workers. Moody's Analytics adjusts the ADP data to match those from BLS.

sector unstructured data include text data from social networks (Facebook, Twitter, etc.), pictures (Instagram, etc.), videos (YouTube, surveillance cameras, etc.), satellite images, traffic webcams, data from drones, etc. These data are often the most difficult data to scrub and transform for statistical purposes as they require complicated transformations based on the specific data source.

Overall, large amounts of high-dimensional data resources are held in the private sector by firms that are themselves information-based enterprises. This observation leads to issues of access for federal statistical purposes, which we address later in this chapter and further in Chapter 6. The table also makes clear that the new data resources arise not from the design of a statistician, but as part of other processes. Sometimes the processes generating the data produce information that may be relevant to official statistics, but this is not their primary purpose. Hence, although the data have been collected by these enterprises, they are not, for the most part, immediately usable for statistical purposes or analysis. For some, much processing work would have to be done to create structured numeric data that have statistical utility. Finally, because the data were not designed for a statistical purpose, they tend to be rather lean, that is, not consisting of a large number of attributes describing the measurement unit (e.g., a person or company). Instead, they measure only what is needed by the process producing them for the firm or other entity. Hence, there is a need to blend these new data resources with traditional survey data in new statistical analyses if they are to be used to improve any existing official statistics. Although blending data sources holds the potential to improve federal statistics, there is no guarantee that it will do so; thus, careful evaluation of data sources is necessary (see below).

USING PRIVATE-SECTOR DATA SOURCES FOR STATISTICS

The potential opportunities to use new data resources for building national statistics have been recognized by many countries of the world with the creation of the U.N. Working Group on Big Data[4] in March 2014. The working group acknowledges that "using Big data for official statistics is an obligation for the statistical community based on the Fundamental Principles [of Official Statistics (see Box 2-1)] to meet the expectation of

---

[4]The full members of the working group are Australia, Bangladesh, Cameroon, China, Colombia, Denmark, Egypt, Indonesia, Italy, Mexico, Morocco, Netherlands, Oman, Pakistan, Philippines, Tanzania, the United Arab Emirates, the United States, the U.N. Economic Commission for Europe, the U.N. Economic and Social Commission for Asia and the Pacific, the U.N. Global Pulse, the International Telecommunications Union, the Organization for Economic Cooperation and Development (OECD), the World Bank, and the Statistical Centre for the Cooperation Council for the Arab Countries of the Gulf.

society for enhanced products and improved and more efficient ways of working" (U.N. Economic and Social Council, 2014, p. 1). The goal of the group is to find promising uses of such data for official statistics, specifically focusing on uses for GPS devices, automated teller machines, scanning devices, sensors, mobile phones, satellites, and social media. The working group has created principles for access to big data sources to ensure fair treatment of businesses supplying these data.

To assess how national statistical offices are seeking to use these new data sources, the U.N. Statistical Commission (UNSC) conducted a survey of 93 national statistical offices. The national statistical offices of countries similar to the United States[5] were most interested in using big data for "faster, more timely statistics" (88%), "reducing response burden" (75%), and creating "new products and services" (72%). These new products and more timely statistics were more important than other factors for the use of big data, such as "modernization of the statistical production process" (69%) and cost reduction (63%) (U.N. Economic and Social Council, 2016). Although many countries are interested in various big data sources for official statistics, very few have yet been able to actually produce official statistics based on these sources.

Academic and private-sector organizations have created statistics based on web-scraped data from e-commerce sites such as the Billion Prices Project (see Box 4-1). The project uses prices of products on the Internet to create a daily Consumer Price Index (CPI) for 22 countries (Cavallo and Rigobon, 2016).[6]

Statistics Netherlands has been able to use big data sources to create national statistics. It has drawn on data from road sensors for transportation and traffic statistics (Puts et al., 2016). Due to the large number of sensors detecting vehicles in about 20,000 highway loops, Statistics Netherlands is able to collect around 230 million records a day. The data are anonymous—the sensors do not record identifiable information, such as license plate numbers—but the data do allow for estimates of what kind of vehicle was observed based on the vehicle's length traveling over the sensor, when vehicles entered and exited highways, and the time of day of the observation. After receiving the data and transforming them, the data are

---

[5]The countries in this definition are those that are members of OECD: Australia, Austria, Belgium, Canada, Chile, Czech Republic, Denmark, Estonia, Finland, France, Germany, Greece, Hungary, Iceland, Ireland, Israel, Italy, Japan, Korea, Latvia, Luxembourg, Mexico, Netherlands, New Zealand, Norway, Poland, Portugal, Slovak Republic, Slovenia, Spain, Sweden, Switzerland, Turkey, the United Kingdom, and the United States.

[6]The 22 countries are Argentina, Australia, Brazil, Canada, Chile, China, Colombia, France, Germany, Greece, Ireland, Italy, Japan, Korea, Netherlands, Russia, South Africa, Spain, Turkey, the United Kingdom, the United States, and Uruguay. See http://www.pricestats.com/inflation-series?chart=1836 [November 2016].

BOX 4-1
The Billion Prices Project

The Billion Prices Project was created by Cavallo and Rigobon (2016) at the Massachusetts Institute of Technology with the objective of measuring inflation using online posted prices for goods and services, an approach known as web-scraping. Web-scraping has the ability to transform the data underlying web pages into databases and, through a "data curation" approach, representative prices can be detected. Indeed, the main challenge of using big data is that most of the data are unimportant. Hence, the curation process involves carefully identifying the retailers that will serve as data sources; using web-scraping software to collect the data; then cleaning, homogenizing, categorizing, and finally extracting the information so it can be used in measurement and research applications.

As computing power has become less expensive, data have been downloaded for more than 50 countries and hundreds of retailers worldwide, and the project has constructed daily inflation measures for about 20 countries. The approach is hybrid: part of the information used in the project is collected by BLS (weights and some services such as education and health) to complement the online price data collection.

Data collection using web-scraping is orders of magnitude cheaper than traditional techniques, such as the surveys used to construct the CPI. More than 5 million items are tracked daily in the Billion Prices Project. ZIn contrast, the CPI is based on prices collected on about 80,000 items per month,[a] with the "market basket" of items used by the CPI determined from data collected about consumer spending in the Consumer Expenditure Survey. The CPI is based on both online and offline goods, while the index created by the Billion Prices Project is based on online goods only; however, in categories such as electronics, clothing, hotels, books, entertainment, travel, and food, the dominance of online retailers is imminent. The advantage of the Billion Prices Project is that, even though it does not include all goods, the data are available on a daily basis for a much larger collection of items than is otherwise available.

[a]See Question #8; available: https://www.bls.gov/cpi/cpifaq.htm [November 2016].

cleaned and adjusted for any possible errors—for example, if a sensor was not functioning properly—and estimates are created for the total number of vehicles on the highways. These estimates can be produced extremely quickly if needed. In early 2016, the Netherlands experienced glazed frost, and Statistics Netherlands was able to produce estimates of how the glazed frost had affected traffic within 2 days.[7]

Another example of using high-dimensional data for national statistics comes from a partnership with private-sector mobile phone service pro-

[7]See http://nos.nl/artikel/2079372-helft-minder-verkeer-door-ijzel.html [November 2016].

viders. Ahas and colleagues (2011) created estimates of tourism statistics in Estonia using GPS-based data from mobile phones. Private-sector data sources are also being actively evaluated to provide new indicators of sustainability, especially for developing countries (U.N. Global Pulse, 2016). In fact, economists have used luminosity from satellite images as an estimator of gross domestic product (GDP) in developing countries (Chen and Nordhaus, 2010). Using 1° × 1° grid-cells that examine luminosity could provide important information on such factors as economic output where there is a lack of population or economic statistics, particularly from war-torn countries. However, luminosity has very little value added for countries that have high-quality statistical systems (Chen and Nordhaus, 2010).

Other emerging uses of high-dimensional data combine them with more traditional statistics created by government statistical agencies. Marchetti and colleagues (2015) created estimates of poverty for small areas by blending mobile phone data with other data from the national statistical office in Italy. Statistics Canada (2016b) has begun to use satellite imagery data as an input for agricultural statistics, replacing a survey. Chessa (2016) used retail outlet scanner data to cover a part of the product prices needed for the CPI. The Colombia National Statistics Office (2016) reported blending satellite digital images to improve land use statistics and land coverage statistics. The U.N. Global Pulse (2014) explored using transformed Twitter data to provide real-time food pricing estimates. Daas and Puts (2014) blended social media sentiment data with traditional data sources to measure consumer confidence.

Many big data projects are currently in pilot project phases, including such projects as use by the Australian Bureau of Statistics (ABS) of satellite surface reflectance data to classify crop type and estimate crop production. ABS is still trying to work out many important challenges such as ensuring reliability of the image data over time, aligning data to statistical boundaries, determining proper level of granularity for the data, and identifying the most accurate statistical methods for estimating quantities of interest (Australian Bureau of Statistics, 2015).

In the United States, a number of federal statistical agencies have been exploring and researching private-sector data sources, such as credit card transactions, other information from commercial providers, and information from Internet sources. Some federal statistical agencies are blending private-sector high-dimensional data with traditional data sources. The Bureau of Justice Statistics (BJS) is currently running a pilot project that is web-scraping data from online articles in order to try to improve estimates for arrest-related deaths (see Box 4-2). BLS currently uses scanner data as part of the input for its CPI estimates (see Horrigan, 2013).

Other federal statistical agencies are using private-sector sources to augment information that could be obtained through surveys. The Eco-

BOX 4-2
Web-Scraping to Improve Statistics on Arrest-Related Deaths

BJS has been responsible for reporting arrest-related deaths since the Deaths in Custody Reporting Act (P.L. 106-247) was enacted in 2000. Until recently, BJS relied on state reporting coordinators (SRCs) in the criminal justice system for each state to identify relevant cases of law enforcement homicides. However, some states did not have SRCs who participated in the reporting of this information, and even in the states with participating SRCs, they used varying methodologies to identify arrest-related deaths. In response to these weaknesses in the system, BJS created a pilot program, the Arrest-Related Deaths Program, in March 2015. The program used a hybrid approach of open-source web-scraping along with existing homicide reports. In brief, the system begins by web-scraping for cases and stories in which a suspect dies in police custody and then noting possible causes. The next step is a survey conducted both with law enforcement agencies and medical examiners about each case found from web-scraping. By this process, the program attempts to identify both false positives and false negatives.

The web-scraping process involves several steps. To begin, the pilot program uses a process to filter and sort through a large volume of articles and stories on the web in order to find cases that are in scope of arrest-related deaths. Each night, the scraping will collect 30,000-40,000 different sources of information and news. Exact duplicates—when the same web URL is used in multiple sources and "untrusted domains" that are not the original source for information (e.g., Wikipedia, Reddit, Amazon)—are eliminated. Next, a "text similarity detector" process is performed: using a threshold of 80 percent of the text being the same, duplicated texts are removed. Next, a "relevancy classifier" is performed on the rest of the sources to identify sources relevant to the programs scope. After all these steps, the remaining sources of information are called the "web front end" and constitute about 1,500 of the original 30,000-40,000 cases. Finally, about 10 coders read through these remaining articles and identify and extract information relevant to the program, including date, personal information, and location. This information is then checked—as noted above—by conducting a survey with both law enforcement agencies and medical examiners to confirm the case is in fact an arrest-related death.

The pilot has so far been very successful, identifying more arrest-related deaths than many other open-source collections. Additionally, the program's ability to use surveys to confirm arrest-related deaths with law enforcement agencies and medical examiners has made its estimates less volatile than similar open-source collections. The pilot program has been able to identify more cases than previously captured by BJS: the program identified about 400 arrest-related deaths in its short 3-month trial period in comparison with about 800 per year that the program had received from the coordinators.

nomic Research Service (ERS) has purchased Nielsen and IRI scanner data, which is linked with individual household details, including demographic characteristics of residents, purchases, and prices. The information can be further linked with other geospatial and store characteristic data to get a more complete picture of the food environment for households.

CONCLUSION 4-1 Enormous amounts of private-sector data that are being generated every day have the potential to improve the timeliness and detail of national statistics.

RECOMMENDATION 4-1 Federal statistical agencies should systematically review their statistical portfolios and evaluate the potential benefits of using private-sector data sources.

CHALLENGES TO USING PRIVATE-SECTOR DATA SOURCES FOR FEDERAL STATISTICS

Given the many different data types shown in Table 4-1 (above) and the many different potential private sources for these data, there are similarly a wide range of challenges for agencies seeking to acquire and use those data for federal statistics. Although these data sources hold the potential to add value to official statistics, there are many methodological and logistical issues that would need to be addressed before that potential can be realized. In this section we discuss two of the major challenges—access and quality of the data—and we will explore them more fully in our second report.

Access

The approaches for accessing private-sector data resources are different from those for government-owned data. As noted in Chapter 3, U.S. federal statistical agencies typically develop a bilateral memorandum of understanding or an interagency agreement to codify the terms under which data can be shared between them. However, asking companies to share their data with federal agencies does not start from the same basic trust or common mission that exists among agencies. Although some companies publish their data and allow free access (e.g., Twitter), other companies sell data services and technology platforms. Companies may be reluctant to share their data for several reasons (Groves, 2013): (1) being liable for possible data breaches if their data are linked with government records; (2) increased attention to confidentiality issues and the data private firms have been collecting without much notice from the public; and (3) the possibility that other companies could use their data to create a profitable product, which they would be unable to capitalize on due to the collaborative agreement.

For companies that are willing to provide data, several approaches are possible. As noted above, ERS has purchased Nielsen and IRI scanner data for food policy research. And as described in Chapter 3, the Department of Housing and Urban Development purchased state and local county tax assessment data from Corelogic, which is a private-sector firm that aggregates these data from local sources and sells them to interested parties. Statistical agencies can also form public-private partnerships with private firms to obtain access to their data. Public-private partnerships are defined as a voluntary collaborative agreement between the public and private sectors. These partnerships are distinguished from other forms of public-private cooperation in that the partnership agreements contain defined roles, responsibilities, and rights and are typically characterized by long terms because of the need for longitudinal data (Robin et al., 2016). Data from private companies normally include information from data collection, including either active (survey) or passive (web-scraping) methods; administrative and similar data used for billing customers and targeting services; and transactional data.

Public-private partnerships are typically implemented through long-term contracts. There are four main approaches for access to and use of the data in public-private partnerships: (1) the company providing the data analyzes the data internally and then shares the relevant statistics with the agency; (2) the company transfers the data to the agency for the agency to compute the statistics; (3) the data are transferred to a trusted third party for analysis, and (4) the statistical agency's functions, including data collection and processing, are outsourced to the private firm.

An example of the first type of partnership was used in Mexico where Telefónica analyzed its call detail records in order to assess the effectiveness of public health alerts for the spread of infectious diseases (Frias-Martinez and Frias-Martinez, 2012). Telefónica compared call detail records in the area of a health alert to a hypothetical model where no alert was given for the same area. Thus, by looking at the difference in mobility between the hypothetical model without health alerts and actual mobility with the health alerts, Telefónica was able to gather information about the effectiveness in reduction of infectious diseases due to health alerts, which it subsequently shared with public agencies.

In the second approach listed above, transfer of datasets is a sharing agreement that involves the physical transfer of databases to the statistical agency under a strict protocol that clearly specifies the terms and conditions and includes each party's responsibilities and penalties for not following the agreement. BLS is currently negotiating with some large companies to provide payroll and other internal company data from which BLS will extract relevant information, rather than asking the company to complete its surveys. Although one advantage of this type of agreement is that statistical offices can analyze the data themselves, it is important to note that many agencies may not have the internal capacity to work with private-sector data (Robin et al., 2016).

The third approach listed above is the transfer to a third party to analyze the data from the provider to give to the statistical agency. There is an example of this type of partnership in Estonia, where the national statistical office formed a public-private partnership to create travel statistics based on cell phone call detail records through the analytics company Positium and the central Bank of Estonia, Eesti Pank. Positium has been working with mobile network operators for more than 10 years and has demonstrated that it is a trusted third party between the Estonian national statistical system and the telecom providers. Positium manages important concerns in using the detailed records, including preservation of business secrets, protection of users' privacy, and compliance with privacy legislation. It also offers benefits to the Estonian statistical system, as it possess the technical ability to safely handle data provided by the mobile network operators in its private servers (Robin et al., 2016).

The last approach listed above, outsourcing a statistical agency's functions, can be described as a process in which activities conducted by statistical offices are outsourced to a contractor. This approach is usually adopted for efficiency. It can include traditional collected data as well as nonofficial data sources that are freely available. This approach is quite common for U.S. federal statistical agencies: of the $7.4 billion spent annually on statistical activities across the federal government, approximately $1.5 billion was designated for private contractors in fiscal 2016 (U.S. Office of Management and Budget, 2015b). Often this work involves survey data collection, but it may also include such activities as frame development, sample design, analysis, and report preparation.

Public-private partnerships offer a number of potential benefits to statistical agencies in that they permit access to private data sources, but there are also important risks and challenges in using those sources. Most of the private data provided in some form to statistical offices from public-private partnerships contain important business data about a firm's customers and strategy that could have negative effects for the data provider if accidently released or breached. Privacy and ethical issues are also important to consider in public-private partnerships as data often contain personally identifiable information, which is information that can be used to distinguish or trace an individual's identity, either alone or when combined with other information that is linked or linkable to a specific individual.[8] In addition,

[8]See OMB circular A-130, p. 33; available: https://obamawhitehouse.archives.gov/sites/default/files/omb/assets/OMB/circulars/a130/a130revised.pdf [February 2017].

a firm's customers or clients can be extremely sensitive about other uses of their data. For example, mobile network operators may be concerned that some customers will change providers simply on the basis that mobile network operators are holding their call data records (Infas, 2010).

Finally, incentives and sustainability for both parties need to be considered. Even if there are short-term benefits for both parties, long-run costs may become an issue as new methods of data collection become available that lead to issues of compatibility and completeness for longitudinal datasets. Moreover, statistical agencies may fear becoming dependent on an outside provider who may discontinue providing data at any time or could raise prices when it becomes clear an agency has no other source for the data.

From a company's perspective, there are two primary access issues to consider: the privacy and confidentiality of the data and the profit objective, which come into play in different ways depending on the arrangement between the firm and the statistical agency. If a company has individual credit card data that could be used to assist in the construction of statistical measures, such as GDP or retail sales in the United States, the firm could work with the statistical agency in a couple of different ways with likely different implications. One possibility would be for the private firm and the statistical agency to jointly develop an index, which the company would sell to the statistical office. Privacy concerns would be minimized by providing aggregate statistics to the agency, but there could be implications for potential profits because such an index would also have value to others in the private sector. The statistical office would likely be unable to compensate the private firm sufficiently to keep it from also selling the index to other companies in the private sector.[9]

The second possibility is for a company to sell its raw credit card data to the statistical agency to analyze and combine with the agency's other information. In this approach, the company and the statistical agency could then each develop their own separate indexes, and the company could sell its index to others without necessarily revealing the same information the statistical agency would publish. However, in this case, the firm would be very concerned about risks to the privacy of its clients and losing control of its data.[10] We discuss issues of privacy and data security in detail in the next chapter, but the main point here is that a company's privacy concerns and profit objective collide and make the form of engagement with a sta-

[9]There is also the potential issue of prerelease ownership and access. If early access to the statistics is potentially of value (e.g., to investors), then loss of control over release could be a disadvantage: that is, there could be a risk that the private partner could profit from sharing the statistics before their official release.

[10]Using secure multiparty computing platforms, which we note in Chapter 5 and will discuss further in report 2, may address these concerns.

tistical agency complicated. There is likely no simple solution, but greater engagement between statistical offices and the private sector will be needed to try to meet the challenge.

Data Quality

We began this chapter noting a wide range of domains in which alternative data sources have the potential to contribute to federal statistics, but these sources are not typically simple substitutes for federal surveys, and careful evaluations of quality are needed. Google Flu Trends was designed to predict influenza incidence reports from the Centers for Disease Control and Prevention (CDC), but it represents a cautionary tale in the use of private data sources for national statistics. Although it performed well initially, in early 2013 Google Flu Trends was predicting nearly twice as many doctor visits due to influenza-like illnesses than the actual number of visits collected by the CDC (Lazer et al., 2014) (see Box 4-3). Other examples

BOX 4-3
Google Flu Trends

One of the weaknesses in Google Flu Trends (discussed above) was its dependence on a correlation between entering search terms that *could* be a signal that the user suffered from influenza (e.g., "Achy shoulders, runny nose"). If there were events that affect the nature of that relationship (e.g., media reports of prevention efforts against the flu), such a correlation could change. That is, more people not suffering from the flu may enter such search terms (Lazer et al., 2014).

The initial version of Google Flu Trends used existing data to find the best matches for 50 million search terms to fit only 1,152 data points (Ginsberg et al., 2009), which resulted in including some terms that correlated with random error instead of the underlying relationship (the model was "overfit"), so that many of the search terms that matched the propensity of the flu did not predict actual future cases. After Google Flu Trends updated its methodology in 2009, research showed that Google Flu Trends was not much better than a fairly simple projection using already available, 2-week lagged CDC data (Goel et al., 2010). However, Lazer and colleagues (2014) note that combining the Google Flu Trends data with other health data, such as lagged CDC data, could improve prediction over using either source alone.

More recently, Yang and colleagues (2015) have created a new model called ARGO (AutoRegressive with Google search data) that accounts for changes in people's search behavior over time. The model is able to self-correct by recalibrating every 2 years using search terms and the CDC's historical flu data. The model incorporates seasonal information on flu outbreaks but does not include terms simply related to the winter season.

BOX 4-4
Scanner Data and Economic Indicators

Scanner data come from scanning consumers' sales at retailers. These data usually include goods sold, prices, quantities, and the goods' characteristics. In principle, these data could have tremendous advantages for the construction of several aggregate economic indicators, such as the CPI, retail sales, and economic activity in general. In practice, however, the available data are often incomplete and need to be properly curated.

If every transaction was recorded, then the construction of the CPI would be expected to be more accurate and simpler. The problem is that only rarely are all the details that are needed (i.e., quantities and prices) actually included. Sometimes the retailer aggregates the data by computing the average (daily or weekly) quantity and price, but this procedure implies averaging between several prices: the regular price, the loyalty card price, the sales price, and the discounted price due to coupons. Averaging may also be done across different stores, or retailers may decide to share only a subset of a store's data. Moreover, not all the transactions in the economy are recorded in scanner data, although this problem is relatively minor and likely to be minimal in the future.

Another challenge in using scanner data is that companies that collect scanner data are mostly interested in measures of market share, response of customers to promotions and price changes, impact of advertising, etc. Answering these questions requires data that differ from the data needed to measure the daily sales of each product. That is, scanner data are currently being collected with the marketing, operations, and production set of questions in mind, but a statistical office is interested in measurements of economic activity and aggregate behavior. To satisfy the statistical need, scanner data would need to include a different level of granularity and complete coverage.

have shown how biased data lead to serious problems in prediction models (see Lum and Isaac, 2016).

High-dimensional data sources present a variety of other quality challenges for statistical uses, including coverage of the population and measurement issues. In terms of coverage of the population, there are often concerns about sample bias with these data sources, in part because such data often exist only for the "haves" and not the "have-nots." In addition, social media data on Twitter, for example, are available only for those who choose to use the application (Couper, 2013). Measurement issues also arise with these data sources because, unlike a carefully designed and tested survey question, social media and some other data often are collected without a set stimulus. Similarly, it is difficult to determine how much of a social media post reflects someone's "true" values and beliefs (Couper, 2013). Even seemingly objective and straightforward scanner data can be fraught with measurement issues (see Box 4-4).

There have been some discussions on how to possibly address these issues (see, e.g., Struijs and Daas, 2014). It may be possible to create weights to reduce coverage bias based on the information that users provide in their social media profiles, which can include useful information about age, gender, or social group. However, considerably more research is needed in this area.

> **CONCLUSION 4-2** The data from private-sector sources vary in their fitness for use in national statistics. Systematic research is necessary to evaluate the quality, stability, and reliability of data from each of these alternative data sources currently held by private entities for their intended use.

We discuss fitness for use further in Chapter 6, and we will discuss quality frameworks evaluating fitness for use in our second report. Because of the many sources and potential challenges with private-sector data, as well as the limited resources of the federal statistical agencies, it is necessary that this research be conducted as efficiently and effectively as possible. We note in Chapter 2 that the Interagency Council on Statistical Policy assists OMB in coordinating the federal statistical system. Since this council is composed of the heads of the principal statistical agencies, it is the logical entity, along with OMB, to oversee the development and implementation of such a research agenda by the agencies in a collaborative and complementary manner.

> **RECOMMENDATION 4-2** The Federal Interagency Council on Statistical Policy should urge the study of private-sector data and evaluate both their potential to enhance the quality of statistical products and the risks of their use. Federal statistical agencies should provide annual public reports of these activities.

We provide some additional discussion of data quality issues for alternative data sources in Chapter 6, and the panel will address this issue more deeply in its second report. Although evaluation of specific data sources is best done at the program level, there is a need across the decentralized federal statistical system for greater leveraging of limited resources for research and development of new methods and assessing the quality of data from new sources. Sustainable access to these data sources is fundamental for federal statistical agencies to make progress in evaluating the quality and usefulness of these data sources for federal statistics. Hence, we end this chapter with key questions facing the future use of high-dimensional data for federal statistics:

- Can the United States develop a sustainable mechanism and environment to permit federal statistical agency access to private-sector high-dimensional data for statistical purposes?
- If such access is sustained, how can the quality of these data sources be evaluated for the benefit of all statistical uses of the data?
- If such access is sustained, how can federal statistical agencies detect changes in the data created by the data holders, which may affect statistical estimates?

5

Protecting Privacy and Confidentiality While Providing Access to Data for Research Use

Our discussion so far has largely focused on federal statistical agencies' production of statistics. However, the mission of statistical agencies is not only to produce statistics, but also to provide statistical research access to the data that they collect while protecting the confidentiality of that information. Although federal statistical agencies provide descriptive statistics that are policy relevant (National Research Council, 2013b), data from the federal statistical agencies are also a key resource for policy analysis and evaluation conducted by researchers and evaluators outside of government (National Research Council, 2005). Indeed, for decades U.S. society has profited from applied social science research and policy analysis using federal data, conducted by universities, nonprofit organizations, think tanks, and advocacy groups. These activities are supported by government grants and contracts, private foundations, and corporate and individual donors (National Research Council, 2005).

It is also well recognized that having external users analyze statistical data is key to improving the quality of federal statistical agency processes (Abowd et al., 2004). The importance of microdata access goes beyond the ability to perform primary analyses relevant to policy; it also includes evaluating the data generation process, replicating scientific findings, and building a knowledge infrastructure (Bender et al., 2016).

In this chapter, we first briefly review the work of previous panels of the National Research Council and National Academies of Sciences, Engineering, and Medicine that have examined issues of researcher access to federal data. We then discuss the risks of access and the challenges that statistical agencies face when providing data access and statistics in an ever-changing data environment. Next, we review the common legal foundations for privacy and confidentiality that apply to the federal statistical agencies and federal administrative records: agencies must follow these laws when acquiring information from data providers, in linking different datasets, and in providing access to external researchers. We note some inadequacies of these laws. We summarize a variety of approaches used by federal statistical agencies for providing access to confidential data for statistical purposes, as well as access models from other countries, focusing on those that include combining multiple data sources. We conclude with a brief discussion of some relevant privacy-enhancing technologies, broadly describing their roles in the context of a single dataset (or statistical agency) and the implications for bringing together multiple data sources.

TENSION BETWEEN PRIVATE LIVES AND PUBLIC POLICIES

We note at the outset that the collection and use of personal information by the federal government, including use for statistical purposes, has long raised privacy concerns. Consequently, there are strong protections to guard the privacy of individuals, such as those for the U.S. decennial census, to encourage public participation and to ensure that data are used for the purpose for which they were collected (Allen and Rotenberg, 2016). These protections help ensure that statistical data are not used in ways that could cause adverse impacts on individuals.

Currently, however, the structures of privacy protections for statistical data are under increasing pressure, for several reasons. First, there is increasing use of government statistical data by private organizations that seek to link data collected for statistical purposes with identifiable individuals.[1] This privacy risk arises from public-private partnerships that may lead to data uses not originally anticipated. Second, there are increasing and more sophisticated instances of data breaches and identity theft in the United States. Even data that are gathered for statistical purposes may be subject to misuse by others as a consequence of a data breach.[2]

---

[1]See, for example, http://www.experian.com/marketing-services/insource-demographics.html [December 2016].

[2]In Australia, for example, the Australian Bureau of Statistics (ABS) has had 14 data breaches since 2013, though "none of the notifications related to disclosure of mishandling of any census data, or attempts by an external party to expose or steal information" (see https://www.theguardian.com/australia-news/2016/jul/29/australian-bureau-of-statistics-reports-14-data-breaches-since-2013 [December 2016]). ABS has also faced criticism from a number of privacy and civil liberties groups over changes to the 2016 census that involved the length of retention of Australians' names and addresses. This will mean that for the first time, the census will retain identifiable information on all Australians for 4 years. ABS has said this will allow it to form a "richer and dynamic statistical picture" of the country (see https://www.theguardian.com/technology/2016/jul/25/census-2016-australians-who-dont-complete-form-over-privacy-concerns-face-fines [December 2016]).

The "fair information practices" of the U.S. Department of Health, Education, and Welfare (1973) are the foundation for most modern privacy laws, including the laws that regulate personal information collected by federal agencies. These practices require notice to the individual about information being collected, consent by the individual for the collection, his/her ability to access his/her data, assurance that the data are kept securely, and some enforcement mechanisms for these data protections (see Gellman, 2016).

Since then, there have been many studies that have examined issues of protecting confidentiality and providing data access to researchers. *Private Lives and Public Policies* (National Research Council, 1993, pp. 15-16) provided a key conceptual framework for these issues, clearly defining the tension between privacy and the value of data-informed policy making:

> Private lives are requisite for a free society. To an extent unparalleled in the nation's history, however, private lives are being encroached on by organizations seeking and disseminating information. . . . In a free society public policies come about through the actions of the people. Those public policies influence individual lives at every stage [of life]. . . . Data provided by federal statistical agencies...are the factual base needed for informed public discussion about the direction and implementation of those policies. Further, public policies encompass not only government programs but all those activities that influence the general welfare, whether initiated by the government, business, labor, or not-for-profit organizations. Thus, the effective functioning of a free society requires broad dissemination of statistical information.

Private Lives and Public Policies defined "informational privacy" as "an individual's freedom from excessive intrusion in the quest for information and an individual's ability to choose the extent and circumstances under which his or her beliefs, behaviors, opinions, and attitudes will be shared with or withheld from others" (National Research Council, 1993, p. 22). It is distinguished from "confidentiality," which refers broadly to an obligation not to transmit that information to an unauthorized party. A more modern definition of information privacy addresses the rights and responsibilities associated with the collection and use of personal information (Rotenberg, 2000).

Another aspect to privacy is concerned with what can be inferred about an individual based on the combination of publicly available information sources and the release of statistical information (Fellegi, 1972; Homer et al., 2008) or the taking of publicly observable actions based on statistical information (Calandrino et al., 2011).

There is a series of research efforts to understand how survey respondents think about the privacy and confidentiality of their information and the risks of disclosure, and how these perceptions affect their behavior (e.g., see Singer, 2003; Singer and Couper, 2010; Singer et al., 2003). Federal statistical agencies typically make pledges to survey respondents that they will keep their information confidential and use it only for statistical purposes. Agencies believe that this is fundamental to gaining cooperation from respondents as well as obtaining accurate and complete information (National Research Council, 1979, 2005).

RISKS OF ACCESS

A decade after *Private Lives and Public Policies*, another National Research Council report (2005) noted a significantly increased societal need for data and increased public concern about privacy and confidentiality than had existed in 1993. These trends have continued, if not accelerated, in recent years. High-profile data breaches of companies, health care providers, universities, and federal agencies have raised people's concerns about the security of their information. In particular, the 2015 Office of Personnel Management (OPM) breach of information raised concerns about the government's ability to secure information. According to OPM, the personnel records of 21.5 million current, former, and prospective federal employees and contractors were stolen, including 5.6 million digitized fingerprints that are used as biometric identifiers to confirm identity and user names and passwords that applicants used for their background investigation forms. The Office of Management and Budget (OMB) reported to Congress in 2016 that the U.S. Computer Emergency Readiness Team received notice of 77,183 incidents over the past year[3] and that cyberattacks against federal agencies are likely on the rise.

There are also access problems other than breaches. One example comes from Australia, which experienced a temporary denial of service shutdown of the website for the 2016 Australian census (Ramzy, 2016). Even prior to this event, the 2016 census had generated a great deal of concern about the privacy of census records (Chirgwin, 2016; Warren, 2016) because ABS had decided to retain names and addresses with the data records for 4 years instead of the 18 months retention used with previous censuses, in order to match census data with other sources. The privacy concerns intensified after reports that the census website had been cyberattacked (Ramzy, 2016). And as noted above, there have been multiple breaches of ABS since 2013; these kinds of threats can undermine the entire social science research enterprise itself, resulting in lower voluntary participation to surveys (National Research Council, 2005, 2013a).

[3]See https://obamawhitehouse.archives.gov/sites/default/files/omb/assets/egov_docs/final_fy_2015_fisma_report_to_congress_03_18_2016.pdf [February 2017].

> **CONCLUSION 5-1** Data breaches and identity theft pose risks to the public.

De-identified individual-level microdata files are of great benefit to researchers because they permit a wide variety of detailed analyses. However, they also present a risk to privacy because of the likelihood that combinations of many characteristics would uniquely identify an individual or organization. Survey data linked with administrative data offer even greater benefits to researchers, but they present even greater risks because they bring together more information than is found in a single source.

> **CONCLUSION 5-2** Combining multiple data sources increases risks to the public from data breaches and identity theft.

The proliferation of publicly accessible data, outside of the statistical agencies, has dramatically increased the risks inherent in releasing microdata because these other data sources can be used to re-identify putatively anonymized data. For example, a medical record containing a randomly generated identifier (in place of a name) and an individual's date of birth might be able to be linked with a record from a different source that contains the random identifier and the individual's gender and zip code.[4]

Similarly, supposedly "anonymized" Netflix rating records were re-identified by linkages to Internet Movie Database (IMDb) reviews on the basis of titles and approximate dates of as few as three movies (Narayanan and Shmatikov, 2008). Because these kinds of linkages occur when the de-identification systems are properly implemented, and require no breaches or security violation, it is difficult to know how frequently they occur. In the words of the President's Council of Advisors on Science and Technology (2014, p. xi):

> Anonymization is increasingly easily defeated by the very techniques that are being developed for many legitimate applications of big data. In general, as the size and diversity of available data grows, the likelihood of being able to re-identify individuals (that is, re-associate their records with their names) grows substantially.

The assumption underlying the use of attempted anonymization for privacy-preserving data analysis is that the data analyst can learn nothing about an individual but still learns the desired statistics. This assumption

---

[4]This was the approach used to re-identify the anonymized medical encounter data of Massachusetts Governor William Weld, which linked birthday, zip code, and gender fields in the voter registration records (Sweeney, 1997).

is problematic at best. One idea to overcome this situation is to provide an analyst only with "statistical" access to the data, without direct access to the raw (anonymized or not anonymized) data. Although such an approach has promise, it is not in itself protective of privacy. There are at least two kinds of attacks that could be launched against systems of this sort: reconstruction and membership.

In a reconstruction attack, an analyst can learn the value of a confidential attribute (e.g., suffers from depression) of almost every member of a dataset (or a targeted subpopulation of the dataset) by combining even only relatively accurate estimates of the fraction of members in sufficiently many random subsets of the dataset (or a targeted subpopulation) having the given attribute (Dinur and Nissim, 2003; Dwork et al., 2007; Kasiviswanathan et al., 2013; Muthukrishnan and Nikolov, 2012).

In a membership attack, an analyst could, say, learn whether a target individual is a member of a case group (e.g., people diagnosed with a particular disease) in a genome-wide association study. It would require only the DNA of the target individual (easily obtainable, as a person sheds DNA on everything) together with (approximate) allele (protein) frequency statistics for the case group and the control group (Dwork et al., 2015a; Homer et al., 2008). That is, the released information is merely a set of statistics revealing the approximate frequencies of "C" and "T," or of "G" and "A," in the case and control groups, for a large number of locations in the DNA. In response to the work of Homer and colleagues (2008), the National Institutes of Health prohibited publication of allele frequency statistics in the studies it funds.

A report of the Institute of Medicine (2009, p. 100) considered the general problem of how to enable researchers' use of medical data while safeguarding privacy:

> In recent years, a number of techniques have been proposed for modifying or transforming data in such a way so as to preserve privacy while statistically analyzing the data (reviewed in Aggarwal and Yu, 2008; NRC, 2000, 2005, 2007b,c). Typically, such methods reduce the granularity of representation in order to protect confidentiality. There is, however, a natural trade-off between information loss and the confidentiality protection because this reduction in granularity results in diminished accuracy and utility of the data, and methods used in their analysis. Thus, a key issue is to maintain maximum utility of the data without compromising the underlying privacy constraints.

The report also noted (p. 101):

> Precisely how this body of developing methodologies may be effectively used in the types of health research of the sort envisioned in this report

remains an open question and this is an area of active research. Thus, alternative mechanisms for data protection going beyond the removal of obvious identifiers and the application of limited modifications of data elements are required. These mechanisms need to be backed up by legal penalties and sanctions.

LEGAL FOUNDATION FOR PRIVACY AND CONFIDENTIALITY

Much of the relevant law that governs federal agencies regarding maintenance of information about individuals derives from the fair information practices (U.S. Department of Health, Education, and Welfare, 1973), which set out the rights and responsibilities associated with the collection and use of personally identifiable information. Among the rights of individuals is the ability to know what personal information about them is collected, how it is used, and who has access to it. Among the responsibilities of those who collect and use personal information are the obligations to ensure the data are used for their intended purpose, are timely and accurate, and are protected against unauthorized use or disclosure.

The Privacy Act of 1974 (5 U.S.C. § 552a) established limitations on the use of a person's Social Security number (SSN), which was viewed as the primary key attribute to combine databases. Section 7 of the Privacy Act provides that any agency requesting disclosure of an SSN must "inform that individual whether that disclosure is mandatory or voluntary, by what statutory authority such number is solicited, and what uses will be made of it." Congress recognized the privacy interest and the dangers of widespread use of SSNs as universal identifiers by making unlawful any denial of a right, benefit, or privilege by a government agency because of an individual's refusal to disclose her or his Social Security number (5 U.S.C. § 552a). The relevant Senate committee stated that the widespread use of SSNs as universal identifiers in the public and private sectors is "one of the most serious manifestations of privacy concerns in the Nation."[5] Short of prohibiting the use of SSNs, the provision in the Privacy Act attempts to limit the use of the number to only those purposes where there is clear legal authority to collect an SSN.

In addition to its provisions for the basic protection of individuals' records, the Privacy Act includes provisions that pertain to the use of records for statistical purposes. The law sought to enable the continued use of data generated by federal agencies while safeguarding privacy (Allen and Rotenberg, 2016).

There are exceptions to Privacy Act obligations for the use of fed-

[5]S. Rep. No. 1183, 93d Cong., 2d Sess., reprinted in 1974 U.S. Code Cong. & Admin. News 6916, 6943.

eral agency data for solely statistical purposes. Records may be matched between federal agencies when the matches "produce aggregate statistical data without any personal identifiers" (U.S.C. 552a(a)(8)(B)(i)). Matches may also be performed "to support any research or statistical project, the specific data of which may not be used to make decisions concerning the rights, benefits, or privileges of specific individuals" (5 U.S.C. 552a(a)(8)(B)(ii)). Agencies are also permitted to disclose records "to a recipient who has provided the agency with advance adequate written assurance that the record will be used solely as a statistical research or reporting record, and the record is to be transferred in a form that is not individually identifiable" (5 U.S.C. 552a(b)(5)).

Agencies are permitted to create exemptions to Privacy Act obligations that would otherwise apply if the records are "required by statute to be maintained and used solely as statistical records" (5 U.S.C. 552a(k)(4)). However, the definition of statistical data in the act is narrow: a statistical record means "a record in a system of records maintained for statistical research or reporting purposes only and not used in whole or in part in making any determination about an identifiable individual" (5 U.S.C. 552a(6)).

CONCLUSION 5-3 Privacy laws have established clear limitations on the collection and use of personally identifiable information for statistical purposes. There are also limits on the use of identifiers, such as Social Security numbers, that enable the linkage of distinct record systems. These laws reflect concerns about the use of personal data gathered by federal agencies.

When federal statistical agencies collect survey data from respondents, they usually pledge to keep the information they collect confidential and to use it only for statistical purposes.[6] Statistical agencies are able to make this pledge to respondents because of authority in their authorizing statutes (e.g., Census Bureau's Title 13) or through the Confidential Information Protection and Statistical Efficiency Act of 2002 (CIPSEA). Prior to the passage of CIPSEA (P.L. 107-347), there was a patchwork of legislation protecting the statistical information collected by various federal statistical agencies (National Research Council, 1993; Wallman and Harris-Kojetin, 2004), with some agencies, such as the Census Bureau, having very strong legal protections for the confidentiality of the data they collected, and other

[6]There are some notable exceptions, such as the Census of Governments, which collects public information from state and local governments, and this information is published in identifiable form. Similarly, the National Center for Education Statistics does not pledge to keep information in the Common Core of Data confidential as this basic information about K-12 schools is considered public and is widely distributed and used by the U.S. Department of Education and others.

agencies having no statutory authority to protect the confidentiality of the data they collected for statistical purposes.

CIPSEA established uniform confidentiality protections for information acquired by agencies—including both the principal statistical agencies and recognized statistical units—under a pledge of confidentiality and for exclusively statistical purposes. CIPSEA requires that such information be used exclusively for statistical purposes and not be disclosed by an agency in identifiable form, for any use other than an exclusively statistical use, without informed consent. CIPSEA defines a statistical purpose as "the description, estimation, or analysis of the characteristics of groups, without identifying the individuals or organizations that comprise such groups" (§502(9)(A)) and explains that the definition includes "the development, implementation, or maintenance of methods, technical or administrative procedures, or information resources that support such purposes" (§502(9)(B)).

CIPSEA provides a high and uniform floor of legal protections and includes criminal penalties for disclosure and unauthorized uses of the information. Specifically, intentional disclosure of confidential information is a class E felony punishable by a fine of $250,000 or 5 years in prison or both (see U.S. Office of Management and Budget, 2007). In conjunction with the authorizing statutes for the federal statistical agencies, CIPSEA provides the foundation for acquiring and protecting data not only from survey respondents, but also from administrative agencies and other data providers.

Federal statistical agencies are required to report to OMB on an annual basis on their implementation of CIPSEA and compliance with the requirements in OMB's implementation guidance (see U.S. Office of Management and Budget, 2007). Agencies are required to use a uniform CIPSEA pledge for all of their data collections that are covered by CIPSEA. These pledges must state that the information will be kept confidential, will be used only for statistical purposes, and will not be disclosed to anyone except agency employees and their agents without the data provider's consent. The pledges also state the penalties noted above for willful disclosure. Agencies are also required to annually train and certify that all employees and agents with access to data covered by CIPSEA have completed CIPSEA training and that all statistical products have been reviewed to ensure that there is no disclosure of identifiable information.

CIPSEA permits recognized federal statistical agencies or units[7] to designate external researchers to obtain access to confidential statistical data for exclusively statistical purposes by giving these agencies the

[7]For a list of OMB-recognized statistical agencies and units, see https://obamawhitehouse.archives.gov/omb/inforeg_statpolicy/bb-principal-statistical-agencies-recognized-units [February 2017].

authority to make such researchers their designated agents and to bind them to the same restrictions in their use of the data and the same criminal penalties for disclosure and misuse as the agencies' employees. This authority has enabled greater opportunities for access and analysis of federal statistical data by agencies that did not previously have this authority.

Recently, federal statistical agencies were concerned that a provision in the Cybersecurity Act of 2015 could be used to undermine the confidentiality protections for the data they collect. Specifically, the Federal Cybersecurity Enhancement Act of 2015 (Title II, Subtitle B of the Cybersecurity Act) gives the secretary of the U.S. Department of Homeland Security (DHS) authority to access any information traveling to or from an agency information system notwithstanding any other law. While federal statistical agencies need and welcome cybersecurity protection from DHS, they were concerned that personally identified data could be accessed and used for purposes unrelated to their agency's mission. Although statistical agencies came to agreement with DHS, they have modified their confidentiality pledges to acknowledge the screening of information by DHS.[8]

APPROACHES TO PROTECTING PRIVACY

Statistical agencies are required by law to protect the confidentiality of the data they collect while maximizing their utility. Threats from data breaches and the growing availability of other sources of data that might be used to re-identify individuals or entities require statistical agencies to reconsider how they can maintain data confidentiality. The publication of statistics covering various groups and subgroups requires careful consideration of how to safely release statistical products and of the potential privacy losses that might occur. In this section, we discuss several different approaches to protecting the privacy of data, including minimizing the personal data that are collected, minimizing disclosure risk by restricting the data that are released, controlling access to and use of the data, encrypting data, and using differential privacy techniques to measure and control cumulative privacy loss.

Data Minimization

One approach to data protection in a statistical environment is to minimize or eliminate the collection of personally identifiable information (Agre and Rotenberg, 1998), that is, information that can be used to distinguish or trace an individual's identity, either alone or when combined with other

[8]For example, see *Federal Register,* Vol. 81, No. 235, pp. 88270-88272 (December 7, 2016).

information that is linked or linkable to a specific individual.[9] The concept can be understood in multiple dimensions. First, in many areas of government statistics, there is little or no personal identifiable information gathered (Solove and Schwartz, 2015). For example, the National Oceanic and Atmospheric Administration (NOAA) collects vast amounts of data that are provided in statistical formats to aid government planning and private-sector activity. The data include weather forecasts, hurricane warnings, and climate trends. Other such data include NOAA's recreational fisheries statistics, which provide catch estimates across various regions in the United States, for time periods ranging from monthly to annually. The data from these queries are used by state, regional, and federal fisheries scientists and managers to maintain healthy and sustainable fish stock.

However, many government datasets do involve the collection of information about individuals, and many of these are valuable for policy analysis. For these record systems, agencies need to consider whether it is necessary to include sensitive data elements that may have an adverse effect on individuals if disclosed. Broadly understood, the aim is data minimization, a concept central to modern privacy law (Allen and Rotenberg, 2016). Educational records pose unique challenges because of the interest in longitudinal studies that may result in tracking individuals over their lifetimes, from the educational environment through employment. The more information available on a single individual, the more likely a single data record can be used to re-identify the person. Details about an individual's education and work history could harm the person if used inappropriately. As a consequence, strong methods need to be employed to ensure that statistical data cannot be readily used for re-identifying any individual or otherwise compromising privacy, such as learning sensitive attributes about an individual (these are not the same thing because sensitive attributes can be learned without re-identification). By minimizing the amount of data collected to those with pre-specified necessary uses, re-identification possibilities are reduced.

Restricted Data

Restricting data includes removing explicit identifiers and applying a variety of statistical disclosure limitation methods to the dataset (see Federal Committee on Statistical Methodology, 2006) to reduce the risk of disclosure. To restrict data, before releasing microdata files statistical agencies remove all obvious identifiers. This approach is not sufficient,

[9]Definition of personally identifiable information as given on page 33: https://obamawhitehouse.archives.gov/sites/default/files/omb/assets/OMB/circulars/a130/a130revised.pdf [February 2017].

however, because some people or entities have characteristics or combinations of characteristics that are rare or unique and make them identifiable. Consequently, there are a variety of statistical disclosure limitation techniques that federal statistical agencies use to reduce the disclosure risk of microdata files. These methods include reducing the amount of information released, perturbing microdata by adding noise, and creating simulated microdata.

The amount of information released is often reduced by recoding variables into categories or fewer categories than were originally collected. For example, occupation might be recoded into 10 high-level categories rather than providing the detailed job titles originally recorded from respondents. Similarly, income is often top-coded (e.g., "more than $250,000") or bottom-coded (e.g., "less than $30,000") to avoid revealing very large or small incomes.

A variety of means are also used to perturb microdata, including swapping, blanking and imputing, data blurring, and a combination of micro agglomeration, substitution, subsampling, and calibration (MASSC). Data swapping or switching is done by matching records that have a high risk of disclosure on a predetermined set of variables and swapping all the other variables. This approach introduces uncertainty as to whether a given record actually reflects the real values. The blank and impute method involves deleting some respondents' values for some variables and replacing them with imputed values. Data blurring and microaggregation involve aggregating values across small sets of respondents for selected variables and then replacing the actual value of some other variables with the average. MASSC is a four-step procedure in which the dataset is partitioned into risk strata, and within these strata values of sensitive variables are swapped, and then records are randomly subsampled within each strata to be retained in the dataset. In the final step, calibration weights are assigned to the retained records to preserve the total weighted counts from the original dataset (for more information, see Federal Committee on Statistical Methodology, 2006).

Another approach is to create a completely synthetic dataset based on the relationships among the variables observed in the confidential data. Such synthetic datasets use statistical models to create microdata records that are plausible predictions of an individual record. In total, the synthetic dataset can reproduce many of the statistical conclusions available from the actual dataset. For example, the Survey of Income and Program Participation created a "synthetic beta" by applying multiple imputation techniques to the data after they were linked to earnings data from the Social Security Administration (Benedetto at al., 2013). We note, however, that privacy is not an automatic consequence of the data being synthetic (we touch on this further below).

These statistical disclosure limitation techniques come with a cost: they decrease the precision of the variables in the dataset and they introduce errors into the dataset, which can affect estimates of population parameters, as well as relationships among variables. As discussed below, this is unavoidable. Moreover, these techniques are not equipped with a single, unifying, measure of privacy loss that can be tracked across multiple applications to yield meaningful statements about cumulative privacy loss as the data are used and reused.

Restricted Access

Restricting access uses administrative procedures and technology to restrict who can access the dataset and what kinds of analyses can be done with the data to reduce the risk of disclosure. Federal statistical agencies have also used a number of different modes for researchers to access and analyze "restricted use" datasets. These methods include licensing agreements, remote access, and online data analysis systems (see Federal Committee on Statistical Methodology, 2006).

Online data analysis systems are currently available for some statistical agency datasets. For example, the National Center for Education Statistics (NCES) has three online data query systems: the Data Analysis System (DAS), the National Assessment of Educational Progress (NAEP) Data Explorer, and QuickStats. These systems are able to provide a user with tabulations and correlations matrices and to construct simple weighted least squares or logistic regression models. (QuickStats does not provide modeling tools.) All three of these systems run the queries on data that have already been statistically perturbed. Furthermore, the tabulation output is limited to categorical statistics, weighted counts, and percentages. A formal data use agreement is not necessary to access these systems, but researchers do need to agree that the data will be used only for statistical purposes.

Although statistical agency online analysis systems have grown in sophistication and flexibility in recent years, they do not allow researchers to use popular statistical software, nor do they provide the capability for sophisticated statistical modeling by users for policy analysis. The National Center for Health Statistics (NCHS) has for a number of years used a remote access analysis system[10] whereby researchers send the system a file by way of file transfer protocol with a program, which is scanned for nonallowable commands. The system then attempts to verify that the program is not trying to access unauthorized data files. Assuming there are no detected problems with the scanned files, the program is run on the real

[10]For details, see http://www.cdc.gov/rdc/b2accessmod/acs230.htm [December 2016].

data. After the program is executed, the output is computer-scanned for disclosure problems and, if none are detected, sent back to the researcher within minutes.[11] If there are any issues, an NCHS staff member conducts a manual inspection. If the staff member approves of the output, it is then e-mailed back, usually within a few hours, depending on staff availability.

Licensing

Another way to restrict access is through licensing agreements, which provide more flexibility than online data analysis systems. Licensing agreements allow researchers access to restricted data from their home institution through the use of strict security procedures and legally binding agreements. NCES has used licensing agreements for many years for many of its survey datasets, especially the longitudinal studies.[12]

To obtain an NCES license, a researcher must submit a written proposal that demonstrates the need for the data, as well as affidavits for any and all people who will be working with the data; the license document itself, which includes information concerning the laws that protect data and states penalties for violating terms of the agreement; and a data security plan. The license must be signed by an official from the researcher's institution who can legally bind the institution.[13] Researchers must also agree to unannounced onsite inspections of the research facilities where the data are secured, a review of statistical output before any public release, and to return and destroy the original data along with any derived files at the end of the license. Although some site inspections have uncovered lapses in procedures by individual investigators, there have not been any documented cases of unlawful disclosures of confidential data from the NCES data licensing program.

Federal Statistical Research Data Centers

Another approach for providing access to data for researchers is through federal statistical agency research data centers. Such centers have been used to provide more stringent controls on who has access to data and the conditions under which they have access. A number of statistical agencies have permitted researchers access to their sensitive data only at

[11]For more information on current and planned capabilities, see https://fcsm.sites.usa.gov/files/2016/03/J3_Meyer_2015FCSM.pdf [December 2016].

[12]The Bureau of Labor Statistics uses a similar licensing approach for the National Longitudinal Survey of Youth microdata, and the National Center for Science and Engineering Statistics uses a similar approach for licensing microdata for several of its surveys.

[13]Researchers have to be associated with an academic or other research institution.

a data center located within their headquarters or a regional office.[14] The Census Bureau pioneered research data centers in other locations around the country beginning in the 1980s.

In 2015, the Census Bureau's Research Data Centers were rebranded as the Federal Statistical Research Data Centers (FSRDCs) to reflect the fact that a number of statistical agencies have at least some datasets available through FSRDCs and that there is growing interest in building, sharing, and governing this infrastructure across the statistical agencies. This infrastructure also opens up the possibility of external researchers linking and combining datasets from different statistical agencies, which, in many cases, current statutes do not permit the agencies themselves to do.

FSRDCs are Census Bureau facilities, housed in partner institutions that meet all physical and information security requirements for access to restricted-use micro data of the agencies whose data are accessed at the FSRDCs. There are currently 24 FSRDCs, and they partner with more than 50 research organizations, including universities, nonprofit research institutions, and government agencies.[15]

The FSRDCs provide computing capacity located behind the Census Bureau firewalls to handle large datasets, and researchers can also collaborate with other researchers across the country through that secure computing environment. All FSRDC researchers must obtain Census Bureau special sworn status, which includes a background check and swearing to protect respondent confidentiality for life, and noting that there are significant financial and legal penalties under Title 13 and Title 26 for failure to do so.

Currently, four federal agencies (the Agency for Healthcare Research and Quality, the Census Bureau, the Bureau of Labor Statistics, and the National Center for Health Statistics) directly provide data through FSRDCs, and each agency has its own review and approval process. In addition to the agencies that directly provide their data, nine other agencies that sponsor surveys also participate in the FSRDC program by allowing surveys they cosponsor to be made available. In a further expansion of the role of FSRDCs, administrative data from other federal agencies are also being made more accessible to researchers through them.

Nongovernment Data Enclaves

NORC at the University of Chicago has created a data enclave that provides various data services, including archiving, curating, and indexing the

---

[14]For example, the National Agricultural Statistics Service provides external researchers access to CIPSEA protected microdata for statistical purposes at its data lab in headquarters or at data labs in its 12 regional field offices.

[15]For more information on the centers, see http://www.census.gov/fsrdc [December 2016].

data, as well as statistically protecting confidential information. Researchers are also provided access to analytic data tools to work with in the secured environment. Researchers have the ability to access data both onsite and remotely. Remote access also allows researchers to share and collaborate while working with each other on the data. NORC staff manage the datasets, including education and training of users in order to ensure that the data are appropriately used, disclosed, and kept confidential. Datasets can be provided by federal, state, or local government agencies as well as private firms, universities, foundations, and other institutions. The providing entity sets the parameters for access and use (including linkage) of their datasets, which are administered and implemented by NORC.

The National Agricultural Statistics Service and the Economic Research Service (ERS) jointly sponsor and conduct the Agricultural Resource Management Survey and provide access to these data through the NORC data enclave.[16] NORC staff are designated as agents of the statistical agencies and must adhere to all CIPSEA requirements as agency employees. Researchers submit a research proposal and, if it is approved, they are able to access the data remotely from their worksite. The output is reviewed by an ERS employee for potential disclosures before the researcher is permitted to download it.

Some universities are creating their own data enclaves, which can also house federal statistical data. The Center for Urban Science Progress at New York University is developing a data facility as a secure research setting with datasets, tools, and expert staff to provide research support services to students, faculty, and government employees. The data facility is designed to be user friendly: it includes user authentication and provides services, such as data curation, research project workspace, data access, and database creation. The primary goals of the data facility are to ensure that new and existing data are made available to and used by current and future members of the research community and that both staff in government agencies and local citizens can use the facility in addressing important research problems.

Data Access in Other National Statistical Systems

As noted in Chapter 3, many European countries make more use of administrative records for national statistics than does the United States. They have also created systems to allow access to administrative data for statistical and research purposes (Card et al., 2010).

[16]The available data are from Phases II and III of the Agricultural Resource Management Survey, and the Tenure, Ownership and Transition of Agricultural Land. For procedures to access these data, see https://www.ers.usda.gov/data-products/arms-farm-financial-and-crop-production-practices/contact-us/#RequestAccess [December 2016].

Statistics Denmark provides researchers de-identified data from combined administrative datasets for research projects through a secure server. Researchers and agencies can access administrative data from all government branches, beginning in 1970, on topics including population and demography, labor markets, earnings, income, consumption, prices, general economic statistics, agriculture, manufacturing industries, construction and housing, service sector, transport, environment and energy, external trade, and national accounts and balance of payments. Statistics Denmark manages, combines, and de-identifies information from multiple administrative databases for projects seeking data on the individual, family, household, workplace, and company levels.

Researchers request data through Danish universities and are accepted on the basis of scientific merit. Researchers have to be part of a Danish research environment; foreign researchers have to be affiliated with a Danish authorized environment. If approved, researchers access data remotely through a secure server (Statistics Denmark, 2014). Data can also be easily linked to other data sources, such as survey data or data from other government agencies. In addition, Statistics Denmark can carry out interview surveys customized to subject needs.

Another model for providing access to administrative data for statistical and research uses is the Administrative Data Research Network (ADRN) in the United Kingdom. The ADRN is made up of four data centers, one for each of the countries of the United Kingdom, and each data center is a secure location where researchers are able to request de-identified data sets for economic and social research. The ADRN functions by serving as a data broker that is able to acquire data for research purposes. The main partners that provide data to the ADRN are the UK Statistics Authority, the Economic and Social Research Council, and data custodians (government departments and agencies and national statistics authorities). To access the data for research purposes, the requester of the data must have a noncommercial and feasible goal for the project that provides public benefit, has scientific merit, and is ethically approved by the ADRN.

In addition to acquiring datasets, the ADRN is able to link and de-identify datasets for researchers. Two examples of linkage that the ADRN has done are linking benefits and earning data with health data to learn more about the impact of poverty on health and linking education data with crime data to understand how education affects criminality (Administrative Data Research Network, 2015). However, it is important to note that since the ADRN must request datasets, difficulties have arisen in acquiring some datasets as there is no mandate that requires data be given to the ADRN.

CONCLUSION 5-4 Federal statistical agencies have a strong tradition of confidentiality and data stewardship. There are growing threats to data repositories and personal privacy that need to be addressed to support this tradition.

CONCLUSION 5-5 A continuing challenge for federal statistical agencies is to produce data products that safeguard privacy. This challenge is increased by the use of multiple data sources.

Using Computer Science and Cryptography to Protect Privacy

So far we have approached the issue of privacy and the protection of the confidentiality of data generally from two directions: government efforts to define and protect privacy and confidentiality through legislation and statistical agencies' attempts to balance the need to make data accessible while still respecting privacy and ensuring confidentiality. We now approach the issues from the domains of theoretical computer science and cryptography.

There is a distinction between privacy and security, set out by Turn and Ware (1976, p. 1):

> Privacy is an issue that concerns the computer community in connection with maintaining personal information on individuals in computerized record-keeping systems. It deals with the rights of the individual regarding the collection of information in a record-keeping system about his persons and activities, and the processing dissemination, storage and use of this information in making determinations about him. . . . Computer security includes the procedural and technical measures required (a) to prevent unauthorized access, modification, use, and dissemination of data stored or processed in a computer system, (b) to prevent any deliberate denial of service, and (c) to protect the system in its entirety from physical harm. . . .

Turn and Ware note that privacy and security issues emerged separately in the 1960s until the "privacy cause célèbre," which was the proposal for a National Data Center, intended to be a centralized databank of all personal information collected by federal agencies for statistical purposes. More recently, a report of the President's Council of Advisors on Science and Technology (2014, p. 33) points out:

> Poor cybersecurity is clearly a threat to privacy. Privacy can be breached by failure to enforce confidentiality of data, by failure of identity and authentication processes, or by more complex scenarios such as those compromising availability.

But privacy and security are not equivalent (p. 34):

> When people provide assurance (at some level) that a computer system is secure, they are saying something about applications that are not yet invented: They are asserting that technological design features already in the machine today will prevent such application programs from violating pertinent security policies in that machine, even tomorrow. Assurances about privacy are much more precarious. Since not-yet-invented applications will have access to not-yet-imagined new sources of data, as well as to not-yet-discovered powerful algorithms, it [is] much harder to provide, today, technological safeguards against a new route to violation of privacy tomorrow. Security deals with tomorrow's threats against today's platforms. That is hard enough. But privacy deals with tomorrow's threats against tomorrow's platforms, since those "platforms" comprise not just hardware and software, but also new kinds of data and new algorithms.

We distinguish two scenarios relevant to the discussion of bringing together multiple data sources, sharing and cooperating. For simplicity, we use "dataset" to describe the data held by one organization, although of course in reality a given organization holds many datasets. The key point is that a party has full access to its own dataset. In the sharing scenario, two or more parties (e.g., statistical agencies) pool their data so that all parties have access to all of the data. In the cooperating scenario, the multiple parties agree to cooperate in a computation on the combination of their multiple datasets, but that is the extent of the collaboration. That is, entity A should learn no more about the datasets of entities B and C than can be learned from the result of the computation. The sharing scenario is the subject of the field of secure multiparty computation, studied in the cryptographic literature since the late 1980s (see, e.g., Goldreich et al., 1987). Both the sharing and the collaborating scenarios could be used when combining data from different sources.

Data could be encrypted using state-of-the-art technology both in transit and at its destination to provide protection against harm in the case of data breaches or inappropriate data access. This can be done using mature technology. Surprisingly, advances in cryptography have shown encryption not to be an insurmountable impediment to data utility. Fully homomorphic encryption schemes (Gentry, 2009) permit arbitrary computations on encrypted data, with no need to decrypt anything except the outputs. In functional encryption, a user operating on encrypted data is given a special "key" that will allow the user to learn only the result of a specific computation (Boneh et al., 2011; Sahai and Waters, 2005). We note, however, that secure multiparty computation, fully homomorphic encryption, and functional encryption are not yet mature technologies.

Privacy-Protective Data Analysis

We now turn to privacy concerns that are independent of security and encryption, that is, problems that arise even when all the encryption and security technologies operate perfectly: threats to privacy that come from the desired outputs of statistical data analysis systems. Protection against these threats is the goal of privacy-protective data analysis.

Two important lessons have been learned from the past 15 years of research on confidential data. First, there are fundamental mathematical limits on "how much" can be computed while maintaining any reasonable notion of privacy: extremely detailed estimates of too many statistics can effectively result in a complete loss of privacy (Dinur and Nissim, 2003; Dwork et al., 2007, 2015a; Homer et al., 2008; Kasiviswanathan et al., 2013; Muthukrishnan and Nikolov, 2012). This body of work has come to be called the fundamental law of information recovery (for a review, see Dwork et al., 2017). This law holds even if all data security, encryption, access control, authorization protocols, password protection, network security, data enclave protocols, and programs are working perfectly.

Second, there are mathematical and algorithmic tools to formally quantify and control privacy loss; in some cases, these tools yield the best possible tradeoffs subject to the fundamental limit. There is hope that these tools, or new approaches yet to be invented, can match the fundamental limit in all cases. This is an extremely active area of research.

In other words, together, these findings delineate the tradeoff between the information that one gains through statistical analysis of a dataset and the loss of privacy that can result from those analyses. As statistical information is extracted from a dataset, there is increasing risk of disclosure of individuals in the dataset. This cumulative privacy loss can be conceptualized as a "privacy loss budget": when a specified level of cumulative risk has been attained, the privacy loss budget would have been fully expended. Using a privacy loss budget means acknowledging that increased accuracy must come at the social cost of increased privacy loss. Conversely, to limit privacy loss to a budgeted total, controls or limits must be placed on analysis. This approach would raise a host of implications, such as prioritizing data usage. Who should be given the right and responsibility of setting a privacy loss budget for a given dataset? Who should be given the first choice of statistical analysis? To date, there is no developed social policy for these questions (Abowd and Schmutte, 2016); there is no technical panacea and no mathematical or computer science substitute for what are ultimately issues of judgment. We discuss these implications further below, and we will elaborate on them in our second report.

Success in privacy-preserving data analysis, however, does not obviate the need for strong encryption and other conventional cybersecurity measures. All of these problems arise in the context of particular datasets. A data warehouse exacerbates security concerns by providing a more valuable target whose compromise is more devastating than the compromise of a single source. However, it may improve the situation with respect to privacy-preserving data analysis both for technical reasons (see Dwork et al., 2012) and because it allows for better coordination in decision making about the allocation of the data resource.

The most heavily studied approach to privacy-protective data analysis is differential privacy, which is a definition of privacy tailored to statistical analysis of large datasets, together with a set of algorithmic techniques for carrying out statistical analyses while adhering to the definition (Dwork, 2006; Dwork and Roth, 2014; Dwork et al., 2006, 2015b). Differential privacy is a promise, with specified levels of assurance, that an individual described by a data record in a dataset will not be affected, adversely or otherwise, by allowing that person's data to be used in any study or analysis, no matter what other computational techniques, studies, datasets, or information sources are or become available. At their best, differentially private algorithms can make confidential data widely available for useful data analysis, without resorting to data clean rooms, data usage agreements, data protection plans, or restricted use enclaves. It permits the measurement and control of privacy loss that accumulates over multiple analyses.

Differential privacy can also be defined as the probability that any observed output is essentially unchanged, independent of whether any individual opts into or opts out of a dataset. The probabilities are taken over random choices made by the data analysis algorithm; "essentially" is quantified in the precise privacy loss guarantee. This simple requirement has many powerful consequences. First, it provides a formal measure of privacy loss. This measure allows one to track privacy loss as it accumulates over multiple computations. It also allows the construction of complex algorithms from simple differentially private building blocks (much as a complex program is the combination of simpler subroutines) while tracking and controlling the privacy loss measure. Finally, any output of a differentially private analysis is "future-proof," meaning that it is robust to all algorithmic attacks and information resources that do not yet exist.

Of course, differential privacy cannot be a panacea—the fundamental law can no more be circumvented than can the laws of physics. Moreover, as noted above for other statistical disclosure limitation techniques, differential privacy introduces error—the fundamental law shows that this also holds true. Sometimes differential privacy introduces no more error than the fundamental law shows to be necessary. In other cases, there is a gap between what is currently known in differential privacy and the known minimal amounts of noise. It is possible that the gaps will be closed by further research or that another, not-yet-invented, technology can help in these cases.

Using such formal privacy guarantees requires a new skill set, foreign to most statistical agencies, social science researchers, and data scientists. Synthetic data generated in a differentially private fashion can address these concerns (in general, privacy is not an automatic consequence of the data being synthetic, but a consequence of the method by which the synthetic data are produced). Differentially private synthetic data may be queried in an ad lib manner, with no risk of further privacy loss beyond that incurred in generating the synthetic data. The U.S. Census Bureau uses synthetic data generated with a variant of differential privacy in the agency's OnTheMap tool, which provides aggregate information about where people work and where workers live (see Machanavajjhala et al., 2008). A drawback is that the task of generating synthetic data with rigorous privacy-protective guarantees can require excessive computational resources; moreover, as is always the case with synthetic data, the synthetic dataset has known properties only for the estimates of the specific statistics it has been designed to capture. If an analyst wants to ask a different question, there is no assurance that the estimates would have the needed properties.

A similar problem occurs when new data are incorporated. For example, consider the commonly used example of randomized responses with a 50-50 prevalence of "yes" for the not-sensitive question and a 50-50 randomization to the sensitive or not-sensitive question. A "yes" answer produces a 3:1 Bayes factor for the unobserved true state being "yes" (rather than "no"). However, if the same person is engaged a second time (with another randomization) and again answers "yes," the Bayes factor is now 9:1. A more likely scenario is that the respondent is asked about a different behavior or attitude, using randomized response approach. A second "yes" produces a Bayes factor of between 3:1 and 9:1 for at least one underlying "yes" rather than "no"-"no," with the value depending on the underlying association of the two behaviors or attitudes. Bounding the association gives a bound on the Bayes factor. One needs to keep in mind that this threat to privacy operates when the amount of linked data is broadened beyond that used to develop a privacy budget.

The fundamental law of information recovery makes clear that meaningful privacy guarantees come at a price. Differentially private algorithms are equipped with explicit tradeoffs between privacy and utility. The statistical nature of the utility loss is a property of the algorithm, not the dataset, and as such can be made public with no loss of data privacy. This characteristic can guide a data analyst in interpreting the outputs, much as the margin of error in an opinion poll informs the public of how to understand the reported results.

Some of the newer, more formal, statistical disclosure limitation techniques have been shown to compare well with the more traditional methods described above. Haney and colleagues (2017) state:

> We design private algorithms and show that they have utility comparable to the existing ad-hoc protection system for an establishment-based data product published by the U.S. Census Bureau.[17]

Anonymization techniques face similar challenges. For example, a study on privacy of data from massive open online courses from MITx and HarvardX on the edX platform reported that standard anonymization methods force changes to datasets that threaten replication and extension of baseline analyses (Daries et al., 2014).

Like secure multiparty computation, homomorphic encryption, and functional encryption, differential privacy is not a fully mature technology. Moreover, most statistical agencies do not currently have staff with skills in these techniques. Nonetheless, these technologies are gaining ground: for example, Apple has introduced differential privacy into iOS 10. The Census Bureau is setting up teams to begin to use differentially private methods in its programs. Indeed, Abowd (2016a, pp. 27-28) recently articulated the predicament currently faced by statistical agencies:

> Almost all current disclosure limitation methods used by statistical agencies around the world are based on ad hoc criteria for measuring their effectiveness. They fail the criterion of equal protection under the law because their effectiveness is measured in terms of an agency's best efforts to insure that the ensemble of publications does not violate the confidentiality of any respondents. Those best efforts, while diligently and competently delivered, were predicated on the assumption that most of the information that could be used to compromise the disclosure limitation procedure was inside the agency's firewall. Such an assumption is simply no longer tenable. It must be replaced by assumptions that allow the agency to release the statistical summaries without fear of future attacks. Formally private disclosure limitation procedures meet this condition. And they are really the only player left standing.

CONCLUSION 5-6 As federal statistical agencies move forward with linking multiple datasets, they must simultaneously address quantifying and controlling the risk of privacy loss.

CONCLUSION 5-7 Privacy-enhancing techniques and privacy-preserving statistical data analysis can potentially enable the use of private-sector data sources for federal statistics.

---

[17]See http://tpdp16.cse.buffalo.edu/abstracts/TPDP_2016_3.pdf [December 2016].

RECOMMENDATION 5-1 Statistical agencies should engage in collaborative research with academia and industry to continuously develop new techniques to address potential breaches of the confidentiality of their data.

RECOMMENDATION 5-2 Federal statistical agencies should adopt modern database, cryptography, privacy-preserving, and privacy-enhancing technologies.

As noted above, the fundamental law of information recovery has ramifications for statistical agencies' disclosure limitation activities. Statistical agencies are accustomed to protecting data from individual inappropriate uses and reviewing each statistical product for disclosure risks; they are not accustomed to limiting statistical analysis or prioritizing analyses based on considerations of cumulative privacy loss or of using a privacy loss budget (Abowd, 2016b; Abowd and Schmutte, 2016). For example, how would one decide the privacy loss budget for the Census of Population and Housing and how much of that budget to assign to analyses for legally required redistricting activities, production of statistical summary information, and microdata analyses for general social science investigations. These kinds of questions are not the domain of the statistical experts inside the agencies, nor of those who create the privacy-preserving analysis systems. These policy issues will need to be confronted by the leaders in agencies, the data users, and stakeholders, including respondents and privacy advocates. We will explore these issues further in our second report, but we note here that *answers* to this wide set of issues are beyond the scope of this panel.

6

Advancing the Paradigm of Combining Data Sources

Over the past 10 years democracies around the world have attempted to make government data available to the public, both to increase transparency and to facilitate easier evaluation of government programs. In the United States, the most recent wave of attention by both the Congress and the administration on evidence-based policy making has highlighted access to data as a key issue, as reflected in 2016 law that established the Evidence-Based Policymaking Commission as well as initiatives in the President's budget (see Chapter 3). In many cases, the critical evidence base for evaluation either cannot be assembled or to do so would be extremely time-consuming. Some of these difficulties reflect statutes, regulations, and policies regarding data sharing; some, lack of incentives to change that status quo. Many times these difficulties can be overcome only at a large cost in terms of time and resources. Other times, valuable research questions lie unexamined because of lack of access to the key data.

In this chapter we take a broad perspective on the federal statistical system and the needs of the research community involved in program evaluation studies. Together, they inform the citizenry about the current status of the economy and the well-being of the population and evaluate whether various government actions improve that status. Building on the findings and themes from the previous chapters, we discuss what is needed to facilitate the use of administrative data and other data sources for federal statistics and for research evaluating the efficacy of federal programs.

We believe it is urgent that changes be initiated now because addressing the changes that are needed will take considerable time and effort and will need to include extensive research, upgrades in information technology

(IT) infrastructure, and new skill sets for current and new federal statistical agency staff. Producing legislatively mandated and policy-relevant statistics is costly and requires a considerable time investment, and changes to methods of how those statistics are produced will require new investments. Furthermore, building a new paradigm while continuing to produce critical information for the nation will be difficult, but we believe the alternative of not making fundamental changes now would result in the inability of many statistical programs to meet their core missions and legislative mandates.

As we note in Chapter 2, sample surveys have played a vital role in providing reliable and trustworthy information to inform the public and policy makers. Sample surveys have many virtues, including the ability to measure the precision of the results, design questions tailored to specific data needs, use a variety of data collection modes to best meet the needs and preferences of respondents, and target specific groups of interest. We expect that sample surveys will continue to play an important but not exclusive role in federal statistics (and, more broadly, in social science research).

Federal statistical agencies will need to examine what information is needed to address key public policy issues and then to consider the best way to produce that information. That examination needs to look at what source(s) of data—surveys, administrative data, other sources, or a combination of them—can best meet the information needs. Federal statistical agencies are in the best position to undertake such evaluations and to combine the most useful sources to produce the best statistical estimates possible in a transparent and objective manner.

In the rest of this chapter we first review the current efforts to examine and use administrative records and other new sources of data for federal statistics. We focus particularly on issues of data access and data sharing, including the environment and infrastructure, both legal and physical, that will be needed. Closely tied to these efforts is the needed IT infrastructure and staff technical skills that will be needed to work with some of these new data sources, including processing, cleaning, and editing large volumes of data. We conclude with a discussion of the quality and usability of different data sources for federal statistics and the necessary research and evaluation that is needed both of the data and of the techniques to protect the privacy of the data.

USING ADMINISTRATIVE RECORDS AND OTHER SOURCES OF DATA FOR FEDERAL STATISTICS

Chapters 3 and 4 discuss using government administrative and private-sector data sources to enhance federal statistics. Although it is clear that other data sources are becoming increasingly available, government administrative data have most clearly demonstrated the direct and immediate potential to improve federal statistics. Both inside and outside the United States, administrative data on their own or in combination with sample survey data are being used for the production of high-quality statistics by a wide range of statistical agencies.

The potential for using private-sector data sources to enhance federal statistics is only beginning to be explored, and evaluations of these new sources are not evenly spread across agencies. Much more work is needed and could be done. A recent report of the National Research Council (2014b, p. 123) made the following recommendation:

> **RECOMMENDATION 5:** Under the leadership of the U.S. Office of Management and Budget, the federal statistical system should accelerate (1) research designed to understand the quality of statistics derived from alternative data—including those from social media, other Web-based and digital sources, and administrative records; (2) monitoring of data from a range of private and public sources that have potential to complement or supplement existing measures and surveys; and (3) investigation of methods to integrate public and private data into official statistical products.

The panel endorses this recommendation and notes that it is still relevant today.

Evaluating alternative data sources for federal statistics can best be achieved by the statistical programs with access to other relevant sources of information. However, there is also a need across the decentralized federal statistical system for greater leveraging of limited resources for research and development of new methods, as reflected in the 2014 recommendation.

Individual agency programs have explored various data sources, but there has been little systematic accumulation of knowledge across agencies. As a result, there is no systemwide plan or strategy for a broad examination of private-sector and other alternative data sources to supplement or replace sample surveys. Furthermore, widespread adoption of new IT requirements, quality assessments, and other areas of needed developments has not occurred.

The 2014 National Research Council report anticipated the difficulties in accomplishing this research due to the nature of the highly decentralized federal statistical system (National Research Council, 2014b, p. 123):

> One of the drawbacks of such a system is the lack of a critical mass for the purpose of major research undertakings. The Census Bureau and perhaps the Bureau of Labor Statistics are the only agencies with significant numbers of in-house research staff, although there is exceptional research capability throughout the statistical system. However, many research topics . . . transcend the needs of any one agency and require a more centralized approach if they are to be successfully pursued.

As described in Chapter 3, the panel found clear successes in federal statistical agencies' use of federal administrative data for statistical programs and purposes. And as described in Chapter 4, we also found some promising pilots in exploring and using various private-sector data sources. However, so far these efforts have been fragmented, and fragmented efforts will not be sufficient for the needs of the overall statistical system. There has been a need for systemwide research and development capabilities even as the survey paradigm was evolving; now, with the exploration of new technologies and data sources, that need is even greater (Habermann, 2010). In addition to endorsing Recommendation 5 (above) from the previous report, we note and repeat the recommendations in Chapters 3 and 4 on the need for a systematic approach to the use of new data sources.

> **RECOMMENDATION 3-1** Federal statistical agencies should systematically review their statistical portfolios and evaluate the potential benefits and risks of using administrative data. To this end, federal statistical agencies should create collaborative research programs to address the many challenges in using administrative data for federal statistics.

> **RECOMMENDATION 4-1** Federal statistical agencies should systematically review their statistical portfolios and evaluate the potential benefits of using private-sector data sources.

> **RECOMMENDATION 4-2** The Federal Interagency Council on Statistical Policy should urge the study of private-sector data and evaluate both their potential to enhance the quality of statistical products and the risks of their use. Federal statistical agencies should provide annual public reports of these activities.

While the panel believes that the above recommendations are needed and will benefit the federal statistical system, it also acknowledges the organizational, policy, and legal barriers that prevent collaborative relationships among statistical agencies. It is not clear that sufficient resources currently exist to pursue the kinds of research needed while continuing to produce the statistics that policy makers and the public expect. However, it is equally clear that the status quo is not meeting the research and development needs of the federal statistical system in evaluating new data sources for federal statistics.

DATA ACCESS AND DATA SHARING

As detailed in Chapter 3, federal statistical agencies face obstacles obtaining access to federal administrative data. When the data are held outside the federal government by states, local governments, or private entities, the obstacles are even more daunting. Although recent guidance has encouraged federal agencies' use of administrative data for statistical and program evaluation purposes (U.S. Office of Management and Budget, 2014a), the results have been discrete efforts that have not been cumulative and have not resulted in a standardized process for accessing data across projects or agencies. For the most part, each project involving two or more agencies requires specific memoranda of understanding that are tailored to the project and dataset being used, often specifying exactly which variables from the dataset may be accessed and by whom.

Even when there are no regulatory impediments and both agencies are eager to share data for statistical purposes, those memoranda of understanding often take months of negotiations. In fact, Prell and colleagues (2009) noted that in the life cycle of an administrative data project, the signing of a memorandum of understanding should be considered a midpoint milestone for a project rather than the beginning of the project, because of the extensive time, planning, resources, and effort needed to reach that agreement. The authors also noted that many projects are abandoned before ever attaining this milestone. As we note in Chapter 3, one possible cause of these difficulties is that there is no agency that is directly charged to ensure timely and effective access of program data for statistical purposes.

In an effort to achieve greater objectivity, the evaluation of federal government programs is often conducted by researchers outside the program. However, external, nonfederal researchers face particular hurdles in gaining access to the data that are crucial to an objective evaluation of program efficacy. There is currently no standard procedure for external researchers to access datasets from different agencies for statistical or evaluation research studies. Although statistical agencies provide a variety of secure means to allow researchers access to their data for statistical purposes (see Chapter 5), access to survey microdata or survey data linked to administrative records typically requires submitting a proposal to each agency whose data will be involved in the project. Each agency has its own application and review process for accessing its data.

Acquisition of datasets from states can require considerably more time, sometimes taking more than 2 years to obtain vital records or other state administrative datasets (see, e.g., Lee et al., 2015). The result is that some social science researchers have shifted away from evaluative and empirical research in the United States to studies in other countries that are able to provide more administrative data and to do so much more quickly (Card et al., 2010).

Although the Confidential Information Protection and Statistical Efficiency Act (CIPSEA) provides a common level of legal protection across statistical agencies and sustains the culture of confidentiality protection within the statistical agencies (see Chapter 5), it would need substantial expansion to serve as a sufficient foundation for effective data sharing and access. As detailed in Chapter 3, the Bureau of Economic Analysis, the Bureau of Labor Statistics, and the Census Bureau have not been able to share business data as was explicitly authorized in Subtitle B of CIPSEA (the "statistical efficiency" component) because of a lack of corresponding authorization in the federal tax code. However, even if this specific lack was remedied, the situation would still fail to provide what is needed more broadly for the statistical system to function effectively as a system. Although greater access to tax data would be a key resource that would greatly benefit the quality of data products for other statistical agencies and programs, other sources would also be of benefit (see Chapters 3 and 4). A new paradigm for the system needs to include changes to several laws that prohibit access for statistical purposes or require legal or regulatory changes to permit access for research and statistical purposes.

It is clear that fundamental changes in data access and sharing need to be made for the future of federal statistics and evidence-based policy research. The panel believes that the country can no longer afford the redundancy of individual federal statistical agencies each negotiating on their own with 50 states and the District of Columbia (and, in some cases, other jurisdictions) to access the same dataset for statistical purposes. It is a burden on the states and the agencies that provides no benefits, and it limits the production of useful statistics and research.

The panel believes that the nation needs a secure environment where administrative data can be statistically analyzed, evaluated for quality, and linked to surveys, other administrative datasets, and other data sources. Such an environment would need to have the authority to control access for statistical and research purposes. It would also have to use and continually evaluate and enhance privacy measures. Integration of these efforts into a single entity could achieve many benefits if all statistical agencies could use a secure data-sharing environment. Without a new entity, no scaling of expertise can occur in privacy protection measures, statistical modeling on multiple data sets, and IT architectures for data sharing.

RECOMMENDATION 6-1 A new entity or an existing entity should be designated to facilitate secure access to data for statistical purposes to enhance the quality of federal statistics.

The panel does not recommend a new entity lightly. As we describe throughout this report, however, there are numerous drawbacks to the status quo, so much so that we believe the statistical system is currently hampered in carrying out its mission. There is also tremendous inertia in many parts of the system that will make any changes difficult. We recognize that creation of a new entity will not by itself solve all the problems detailed in this report. In fact, we expect that, like the statistical agencies themselves, the authority and mission of the new entity will need to be clearly delineated, as organizational issues will arise between it and the existing agencies. How this entity is created and its functions will determine its ability to be an effective resource of and for the federal statistical system. Thus, in the remainder of this chapter, we delineate some foundational principles and raise fundamental issues that will need to be addressed in order to create an effective new entity. In our second report, we will explore these issues more deeply.

PROTECTING PRIVACY

As many people in federal statistical and evaluation research communities know, these opportunities and challenges are not new. As Kraus (2013, p. 1) observed, a similar situation occurred in the 1960s:

> Computer technology had improved the efficiency and affordability of research with large data sets, and the expansion of government social programs called for more data and research to inform public policy. As a result, in 1965 social scientists recommended that the federal government develop a national data center that would store and make available to researchers the data collected by various statistical agencies. Because of its massive data holdings and its pioneering work in the use of computers for the storage and analysis of data, the Census Bureau became involved in the national debate, though reluctantly.

However, the proposal for a national data center led to widespread concerns about government profiling and monitoring. An anti-"databank" movement emerged, and there were congressional hearings. The results were an extensive report, *Records, Computers, and the Rights of Citizens* (U.S. Department of Health, Education, and Welfare, 1973), and comprehensive legislation in 1974 that essentially prevented the establishment of a centralized database in the United States. New limitations were adopted for the use of Social Security numbers, understood at the time as the key technique to link discrete record sets containing personally identifiable information. Kraus concluded (2013, p. 1): "One key lesson of the data center debate is that social scientists and government agencies must consider the practical implications of their plans and clearly communicate those plans to the public."

The panel does not envision this new entity as a major new data warehouse or national data center. We will discuss potential IT approaches and requirements in our second report, but emphasize here that there are mechanisms and protocols, such as secure multiparty computing, for combining and analyzing data virtually that do not require all the data being combined to be in the same place. Given the privacy threats that the public already experiences and the history of the proposal for a national data center, it is clear that privacy protections must be at the forefront of the design and administration of a new entity, using technological and administrative approaches to secure the data, along with cutting-edge privacy-preserving and privacy-enhancing techniques. In addition, staff with skills in cryptography and computer science will be needed to research and use new privacy-preserving and privacy-enhancing techniques for survey and linked datasets. It will also be critical that the governance of the panel's proposed entity acknowledges people's right to know how their data are being used, and the concerns of the public must guide the practices of the entity. Transparency and continuously improving privacy protections will need to be the hallmark of the entity as we expect threats to privacy and confidentiality to continuously evolve.

In order to fully take advantage of currently available technology and administrative data sources, it is important that the proposed entity have sufficient staff with technical expertise to remain a functional, improving, and permanent entity. There are also economies of scale to be realized by a centralized entity. It would be impractical and wasteful for each statistical agency to try to attract and maintain the needed technical staff and to provide the IT infrastructure necessary to be able to extract, transform, load, clean, link with survey data they collect, as well as analyze a wide array of new datasets and data streams from federal, state, and local governments and private entities.

The Census Bureau has invested substantial resources into the Center for Administrative Records Research and Applications (CARRA) and has amassed considerable IT infrastructure and technical staff for linking and processing survey and administrative data. Building and maintaining this capacity centrally, for all of the statistical agencies to use, would be much more effective and cost-efficient than attempting to replicate this model across more than a dozen agencies. Small statistical agencies would not be able by themselves to create the infrastructure or attract the people with the needed skills; they need to be able to rely on the overall system to provide this technical assistance.

CONCLUSION 6-1 For the proposed new entity to be sustainable, the data for which it has responsibility would need to have legal protections for confidentiality and be protected, using the strongest privacy protocols offered to personally identifiable information while permitting statistical use.

RECOMMENDATION 6-2 The proposed new entity should maximize the utility of the data for which it is responsible while protecting privacy by using modern database, cryptography, privacy-preserving, and privacy-enhancing technologies.

Extending the recommendation, we offer a set of prerequisites for the successful organization of the proposed new entity:

1. It has to have legal authority to access data that can be useful for statistical purposes. The legal authority needs to span cabinet-level departments and independent agencies.
2. It has to have strong authority to protect the privacy of data that are accessed and prevent misuse. At minimum, that authority needs to be commensurate with existing laws (CIPSEA, the Privacy Act), but it may also require new legislation.
3. It has to have authority to permit appropriate uses for the extraction of statistical information from the multiple datasets relevant to program evaluation and the monitoring of policy-relevant social and economic phenomenon. The authority needs to delimit what uses are forbidden as well as what uses are encouraged.
4. It needs to be staffed with personnel whose skills fit the needs of the proposed entity, including advanced IT architectures, data transmission, record linkage, statistical computing, cryptography, data curation, cybersecurity, and privacy regulations.

Without these features, it is doubtful that a sustainable data-sharing environment could be constructed.

The panel stresses that it views this new entity as collaborative with federal statistical agencies. It should provide a platform for data sharing and enhancement of statistical programs, as well as for facilitating much-needed collaborative research with new sources of data. It should not take over their programs or authorities nor be a drain on federal statistical system resources.

In addition to the necessary features, however, much remains undetermined. The goal is to design an entity that can address the difficulties that statistical agencies have in accessing, evaluating, and using administrative and private-sector data sources for federal statistics. Any new entity will have pros and cons. At this point in our work, the panel has identified six key issues that need to be carefully considered in designing a successful data-sharing environment:

1. Should the entity be located in an existing organization, or should it be a new organization? Since it needs to facilitate new uses of multiple data sources, should it be a newly funded unit in an existing statistical agency? Should it be a new federal unit shared by all federal statistical agencies? Should it be located in a program agency? Should it be a new Federally Funded Research and Development Center to offer more flexibility of staffing? Should it be a new private-government-academic institution, with shared governance? If the entity will be a new organization, will it have its own institutional review board, disclosure review board, privacy officer, and other regulatory attributes of research environments?
2. Should the organization be an environment that permits access to data owned and stored by partnering organizations, storing no data itself, or should it be a data repository? Should the entity be responsible for curating and storing all editions of a given dataset? Should it be responsible for all the metadata for the data that it holds, or should that be the responsibility of the providing organizations?
3. How should access for federal and nonfederal research uses be administered? Will the environment be one in which only outside research staff can access data? For example, the entity could be staffed by data curators and experts in data merging, matching, and dataset construction, with data analysis controlled and directed by federal and nonfederal external researchers. Alternatively, the entity could be a "full-service" research institute, with both internal and external federal researchers having access to data. Nonfederal researchers could affiliate with the entity under appropriate controls.
4. What transparency features should be in place for the entity? Should public notification be made for all uses of data accessible through it?
5. How can the entity best apply state-of-the-art privacy protections? How can it be set up to respond quickly both to new privacy threats and new privacy-protecting research developments?
6. How will the entity be financed? Will there be annual appropriations and, if so, what would be the authorizing source? Other possibilities for funding would be through agreements with federal statistical and program agencies or by charging user fees to the research community.

Each of these issues and questions requires careful consideration. There may be multiple possible answers to the questions above, but any move toward establishing a new entity needs to have at least one feasible answer to each of them. The answers to these questions will help to determine what cooperative efforts between branches of government, and legislation, might be needed.

We note some of the conclusions in other chapters are relevant to the new entity: the conclusions concerning legal barriers to accessing federal administrative data (in Chapter 3) and on the use of state and local administrative data from federally funded programs (Chapter 4) also affect the new agency.

> CONCLUSION 6-2 To carry out its purpose of facilitating secure access to federal program administrative data for statistical purposes, the new entity would need to be able to legally access those data.

> CONCLUSION 6-3 To encourage states and local authorities to provide access to their administrative data for statistical purposes, the new entity would need to have authority to provide incentives to them.

ASSESSING DATA QUALITY AND FITNESS FOR USE

As we have argued throughout this report, the federal statistical system fulfills a vital role for the country by providing high-quality, objective information for the public good and to inform decision making for both the public and private sectors. There are now real opportunities to improve the information infrastructure and federal statistics through greater access and leveraging of government administrative data and other new public and private data sources; however, there are many challenges with using these new sources, and these sources need further exploration and systematic evaluation. The panel recommends in this report that the barriers that impede access to these data sources for federal statistics be removed to enable federal statistical agencies to conduct the careful, systematic research into using those sources, and that they be used only for statistical purposes.

The panel envisions that statistical agencies will systematically evaluate individual data sources for fitness for a specific use, timeliness, consistency (across years and across jurisdictions), completeness, and accuracy. Agencies would then use a combination of data sources, taking advantage of the strengths of each source, to produce key statistics and the data needed for public policy, and they would do so in a transparent manner with documentation of appropriate measures of uncertainty. They would also evaluate the impact of using multiple data sources on the continuity of leading economic and social indicators.

The panel also recognizes much work is needed to achieve what we envision. For example, the area of financial market data is in some ways far more advanced in terms of matching and blending different types of data, and those advances have been aided by the propagation of standardization. Standardized messaging systems allow transactions to proceed on a global basis: for example, the Global Legal Entity Identifier System was created to provide a globally coherent facility for identifying entities.[1] If administrative and other data sources are to be used for federal statistics, standardization will be needed for entities, and standards will be needed for determining when data are "fit" for use. In Chapters 3 and 4 we offer conclusions and recommendations for statistical agencies to conduct further research on the utility and quality of administrative records and other alternative data sources for use in federal statistics. However, no single agency can develop standards for fitness for use; it is a systemwide task and obligation. Indeed, statistical agencies will likely also need to collaborate with academia and industry to do this work.

In our second report we will discuss approaches for implementing a new paradigm that would combine diverse data sources from government and private-sector sources, including further elaboration of the characteristics needed for the proposed new entity, as well as the IT implications. We will discuss the framework needed to evaluate the quality of alternative sources and the estimates that come from combined data, and we will evaluate the concepts, metrics, and methods for assessing the quality and utility of alternative data sources, analogous to the "total error" framework used for surveys. We will also discuss in greater detail the statistical methods for combining multiple data sources, including those for various statistical modeling approaches, small-area estimation, and combining multiple frames. We will also further examine and review current research and approaches for privacy protections. As appropriate in each of these domains, we will provide recommendations for a research agenda.

---

[1]See https://www.gleif.org/en/about-lei/introducing-the-legal-entity-identifier-lei [January 2017].

References

Abowd, J. (2016a). *How Will Statistical Agencies Operate When All Data Are Private?* Available: http://digitalcommons.ilr.cornell.edu/ldi/30 [December 2016].

Abowd, J. (2016b). *Why Statistical Agencies Need to Take Privacy-Loss Budgets Seriously, and What It Means When They Do.* Proceedings of the Federal Committee on Statistical Methodology Policy Seminar, Washington, DC, December 7. Available: http://digitalcommons.ilr.cornell.edu/ldi/32 [December 2016].

Abowd, J., and Schmutte, I. (2016). *Revisiting the Economics of Privacy: Population Statistics and Confidentiality Protection as Public Goods.* Available: http://digitalcommons.ilr.cornell.edu/ldi/22 [December 2016].

Abowd, J., and Vilhuber, L. (2005). The sensitivity of economic statistics to coding errors in personal identifiers. *Journal of Business and Economics Statistics, 23*(2), 133-152.

Abowd, J., Haltiwanger, J., and Lane, J. (2004). Integrated longitudinal employer-employee data for the United States. *American Economic Review Papers and Proceedings, 94*(2), 224-229.

Administrative Data Research Network. (2015). *Better Knowledge Better Society: Network Review.* Available: https://adrn.ac.uk/media/1193/adrn-annual-review-2014-2015_web.pdf [November 2016].

Advisory Commission to Study the Consumer Price Index. (1996). *Toward a More Accurate Measure of the Cost of Living.* Available: https://www.ssa.gov/history/reports/boskinrpt.html [November 2016].

Agre, P.E., and Rotenberg, M. (1998). *Technology and Privacy: The New Landscape.* Cambridge, MA: The MIT Press.

Ahas, R., Tiru, M., Saluveer, E., and Demunter, C. (2011). *Mobile Telephones and Mobile Positioning Data as Source for Statistics: Estonian Experiences.* Available: http://citeseerx.ist.psu.edu/viewdoc/download?doi=10.1.1.461.3362&rep=rep1&type=pdf [November 2016].

Allen, A.L., and Rotenberg, M. (2016). *Privacy Law and Society* (3rd Edition). St. Paul, MN: West Academic Publishing.

Australian Bureau of Statistics. (2015). *Report from the Task Team on Satellite Imagery, Remote Sensing and Geospatial Data.* Available: http://unstats.un.org/unsd/trade/events/2015/abudhabi/presentations/day2/01/1%20ABu%20Dhabi%20-%20Introduction%20and%20overview%20-%2020%20October%20version.pdf [November 2016].

Baker, R., Blumberg, S., Brick, J.M., Couper, M., Courtright, M., Dennis, J.M., Dillman, D., Frankel, M., Garland, P., Groves, R., Kennedy, C., Krosnick, J., and Lavrakas, P. (2010). AAPOR report on online panels. *Public Opinion Quarterly,* 1-71. Available: https://pprg.stanford.edu/wp-content/uploads/2010-AAPOR-Report-on-Online-Panels.pdf [November 2016].

Becker, A. (2015). *Ninety-One Percent of Colleges Report Zero Rapes in 2014.* Available: http://www.aauw.org/article/clery-act-data-analysis [November 2016].

Bellhouse, D.R. (2000). Survey sampling theory over the twentieth century and its relation to computing technology. *Survey Methodology, 26,* 11-20.

Bender, S., Jarmin, R., Kreuter, F., and Lane, J. (2016). Privacy and confidentiality. In I. Foster, R. Ghani, R.S. Jarmin, F. Kreuter, and J. Lane (Eds.), *Big Data and Social Science: A Practical Guide to Methods and Tools* (pp. 299-312). New York: CRC Press.

Benedetto, G., Stinson, M., and Abowd, J.M. (2013). *The Creation and Use of the SIPP Synthetic Beta.* Suitland, MD: U.S Census Bureau. Available: https://www.census.gov/content/dam/Census/programs-surveys/sipp/methodology/SSBdescribe_nontechnical.pdf [November 2016].

Beuken, Y., and Vlag, P. (2010). *Business Register: The Dutch Experience.* Available: https://www.ine.pt/filme_inst/essnet/papers/Session4/Paper4.3.pdf [November 2016].

Blumberg, S.J., and Luke, J.V. (2007). Coverage bias in traditional telephone surveys of low-income and young adults. *Public Opinion Quarterly, 71*(5), 734-749.

Blumberg, S.J., and Luke, J.V. (2011). *Wireless Substitution: Early Release of Estimates from the National Health Interview Survey, January-June 2011.* Available: http://www.cdc.gov/nchs/data/nhis/earlyrelease/wireless201112.pdf [November 2016].

Blumberg, S.J., and Luke, J.V. (2016). *Wireless Substitution: Early Release of Estimates from the National Health Interview Survey, July-December 2015.* Available: http://www.cdc.gov/nchs/data/nhis/earlyrelease/wireless201605.pdf [November 2016].

Blumerman, L.M., and Vidal, P.M. (2009). *Uses of Population and Income Statistics in Federal Funds Distribution—With a Focus on Census Bureau Data.* Governments Division Report Series, Research Reports No. 2009-1. Available: https://www.census.gov/prod/2009pubs/govsrr2009-1.pdf [November 2016].

Boneh, D., Sahai, A., and Waters, B. (2011). *Functional Encryption: Definitions and Challenges.* Proceedings of the Theory of Cryptography Conference (TCC) 2011, Baltimore, MD, November 13-15. Available: https://eprint.iacr.org/2010/543 [November 2016].

Brick, J.M., and Williams, D. (2013). Explaining rising nonresponse rates in cross-sectional surveys. *The ANNALS of the American Academy of Political and Social Science, 645*(1), 36-59.

Bureau of the Census. (1975). *Historical Statistics of the United States: Colonial Times to 1970.* Available: https://www.census.gov/history/pdf/histstats-colonial-1970.pdf [November 2016].

Bureau of Justice Statistics. (2003). *Level of UCR and NIBRS Participation.* Available: http://www.bjs.gov/content/nibrsstatus.cfm [November 2016].

Bureau of Labor Statistics and U.S. Census Bureau. (2006). *Design and Methodology: Current Population Survey.* Technical Paper 66. Available: http://www.census.gov/prod/2006pubs/tp-66.pdf [November 2016].

Calandrino, J.A., Kilzer, A., Narayanan, A., Felten, E., and Shmatikov, V. (2011). *"You Might Also Like": Privacy Risks of Collaborative Filtering.* Available: https://www.cs.utexas.edu/~shmat/shmat_oak11ymal.pdf [November 2016].

California State Auditor. (2015). *California Post Secondary Educational Institutions: More Guidance Is Needed to Increase Compliance With Federal Crime Reporting Requirements*. Report 2015-032. Available: http://auditor.ca.gov/pdfs/reports/2015-032.pdf [November 2016].

Card, D., Chetty, R., Feldstein, M., and Saez, E. (2010). *Expanding Access to Administrative Data for Research in the United States*. Arlington, VA: National Science Foundation. Available: www.nsf.gov/sbe/sbe_2020/submission_detail.cfm?upld_id=112 [November 2016].

Catalano, S.M. (2004). *Crime Victimization, 2003*. Bureau of Justice Statistics, NCJ 205455. Available: https://www.bjs.gov/content/pub/pdf/cv04.pdf [November 2016].

Cavallo, A., and Rigobon, R. (2016). *The Billion Prices Project: Using Online Prices for Measurement and Research*. Working Paper 22111. Cambridge, MA: National Bureau of Economic Research. Available: http://www.nber.org/papers/w22111 [November 2016].

Centers for Disease Control and Prevention. (2015). *National Immunization Survey: A User's Guide for the 2014 Public-Use Data File*. Available: ftp://ftp.cdc.gov/pub/Health_Statistics/NCHS/Dataset_Documentation/NIS/NISPUF14_DUG.pdf [November 2016].

Chen, X., and Nordhaus, W. (2010). *The Value of Luminosity Data as a Proxy for Economic Statistics*. Working Paper 16317. Cambridge, MA: National Bureau of Economic Research. Available: http://www.nber.org/papers/w16317 [November 2016].

Chessa, A.G. (2016). *Processing Scanner Data in the Dutch CPI: A New Methodology and First Experiences*. Proceedings of the Meeting of the Group of Experts on Consumer Price Indices, Geneva, Switzerland, May 2-4. Available: https://www.unece.org/fileadmin/DAM/stats/documents/ece/ces/ge.22/2016/Session_1._Netherlands_Processing_scanner_data_in_the_Dutch_CPI.pdf [December 2016].

Chirgwin, R. (2016). *The Australian Bureau of Statistics Has Made a Hash of the Census*. Available: http://www.theregister.co.uk/2016/08/01/the_abs_has_burned_trust_and_thats_a_problem [November 2016].

Citro, C. (2014). From multiple modes for surveys to multiple data sources for estimates. *Survey Methodology, 40*(2), 137-161.

Cochran, W.G. (1953). *Sampling Techniques*. New York: Wiley; London: Chapman & Hall.

Cohany, S.R., Polivka, A.E., and Rothgeb, J.M. (1994). Revisions in the Current Population Survey effective January 1994. In *Employment and Earnings* (pp. 13-37). Washington, DC: Bureau of Labor Statistics. Available: http://www.bls.gov/cps/revisions1994.pdf [November 2016].

Colombia National Statistics Office. (2016). *Use of Satellite Images to Calculate Statistics on Land Cover and Land Use*. Available: http://cepei.org/wp-content/uploads/2016/08/report-pilot-project-colombia-v3.pdf [November 2016].

Comey, J.B. (2015). *Hard Truths: Law Enforcement and Race*. Available: https://www.fbi.gov/news/speeches/hard-truths-law-enforcement-and-race [November 2016].

Couper, M. (2013). Is the sky falling? New technology, changing media, and the future of surveys. *Survey Research Methods, 7*(3), 145-156.

Cruze, N.B. (2015). Integrating survey data with auxiliary sources of information to estimate crop yields. In *Proceedings of the Survey Research Methods Section* (pp. 565-578). Washington, DC: American Statistical Association.

Czajka, J., and Beyler, A. (2016). *Declining Response Rates in Federal Surveys: Trends and Implications*. Washington, DC: Mathematica Policy Research.

Czajka, J.L., and Denmead, G. (2008). *Income Data for Policy Analysis: A Comparative Assessment of Eight Surveys*. Washington, DC: Mathematica Policy Research.

Daas, P., and Ossen, S. (2011). Metadata quality evaluation of secondary data sources. *International Journal for Quality Research, 5*(2), 57-66.

Daas, P., and Puts, M. (2014). *Social Media Sentiment and Consumer Confidence*. European Central Bank and Statistics Paper Series No. 5. Available: http://www.pietdaas.nl/beta/pubs/pubs/Ecbsp5.pdf [November 2016].

Daas, P.J.H., Puts, M.J., Buelens, B., and van den Hurk, P.A.M. (2015). Big data as a source for official statistics. *Journal of Official Statistics, 31*(2), 249-262.

Daries, J.P., Reich, J., Waldo, J., Young, E.M., Whittinghill, J., Ho, A.D., Seaton, D.T., and Chuang, I. (2014). Privacy, anonymity, and big data in the social sciences. *Communications of the ACM, 57*(9), 56-63.

deLeeuw, E.D. (2008). Choosing the method of data collection. In E.D. deLeeuw, J.J. Hox, and D.A. Dillman (Eds.), *International Handbook of Survey Methodology* (pp. 113-135). New York: Lawrence Erlbaum Associates.

Deming, W.E. (1950). *Some Theory of Sampling*. New York: Dover.

Dinur, I., and Nissim, K. (2003). *Revealing Information while Preserving Privacy*. Available: http://www.cse.psu.edu/~ads22/privacy598/papers/dn03.pdf [November 2016].

Duncan, J.W. (1976). Confidentiality and the future of the U.S. statistical system. *The American Statistician, 30*(2), 54-59.

Duncan, J.W., and Shelton, W.C. (1992). U.S. government contributions to probability sampling and statistical analysis. *Statistical Science, 7*(3), 320-338.

Dwork, C. (2006). Differential privacy. In *Automata, Languages and Programming* (Vol. 4052) (pp. 1-12). Heidelberg, Germany: Springer. Available: http://research.microsoft.com/pubs/64346/dwork.pdf [November 2016].

Dwork, C., and Roth, A. (2014). The algorithmic foundations of differential privacy. *Foundations and Trends in Theoretical Computer Science, 9*(3-4), 211-407.

Dwork, C., McSherry, F., Nissim, K., and Smith, A. (2006). Calibrating noise to sensitivity in private data analysis. In S. Halevi and T. Rabin (Eds.), *Theory of Cryptography* (Vol. 3876) (pp. 265-284). Heidelberg, Germany: Springer. Available: http://link.springer.com/chapter/10.1007/11681878_14 [November 2016].

Dwork, C., McSherry, F., and Talwar, K. (2007). The price of privacy and the limits of LP decoding. In *Proceedings of the 39th ACM Symposium on Theory of Computing* (pp. 85-94). New York: ACM Publications. Available: http://dl.acm.org/citation.cfm?id=1250804 [November 2016].

Dwork, C., Naor, M., and Vadhan, S. (2012). *The Privacy of the Analyst and the Power of the State*. Proceedings of the 2012 IEEE 53rd Annual Symposium on Foundations of Computer Science (FOCS), New Brunswick, NJ, October 20-23. Available: http://privacytools.seas.harvard.edu/files/privacytools/files/06375318.pdf [November 2016].

Dwork, C., Smith, A., Steinke, T., Ullman, J., and Vadhan, S. (2015a). *Robust Traceability from Trace Amounts*. Proceedings of the 2015 IEEE 56th Annual Symposium Foundations of Computer Science (FOCS), Berkeley, CA, October 17-20. Available: http://privacytools.seas.harvard.edu/files/privacytools/files/robust.pdf?m=1445278897 [November 2016].

Dwork, C., Feldman, V., Hardt, M., Pitassi, O., Reingold, O., and Roth, A. (2015b). The reusable holdout: Preserving validity in adaptive data analysis. *Science, 349*(6248), 636-638.

Dwork, C., Smith, A., Steinke, T., and Ullam, J. (2017). Exposed! A survey of attacks on private data. *Annual Review of Statistics and Its Application, 4*.

The Economist. (2012). Don't lie to me, Argentina. Available: http://www.economist.com/node/21548242 [November 2016].

Federal Committee on Statistical Methodology. (2006). *Report on Statistical Disclosure Limitations Methodology*. Statistical Policy Working Paper No. 22. Available: https://fcsm.sites.usa.gov/files/2014/04/spwp22.pdf [November 2016].

Fellegi, I.P. (1972). On the question of statistical confidentiality. *Journal of the American Statistical Association, 67*(337), 7-18.

Freire, J., Bessa, A., Chirigati, F., Vo, H., and Zhao, K. (2016). *Exploring What Not to Clean in Urban Data: A Study Using New York City Taxi Trips*. New York: New York University. Available: http://sites.computer.org/debull/A16june/p63.pdf [November 2016].

Frias-Martinez, V., and Frias-Martinez, E. (2012). *Enhancing Public Policy Decision Making Using Large-Scale Cell Phone Data*. Available: www.unglobalpulse.org/publicpolicyandcellphonedata [November 2016].

Gellman, R. (2016). *Fair Information Practices: A Basic History*. Available: http://papers.ssrn.com/sol3/papers.cfm?abstract_id=2415020 [December 2016].

Gentry, C. (2009). *A Fully Homomorphic Encryption Scheme*. Ph.D. dissertation. Stanford, CA: Stanford University. Available: https://crypto.stanford.edu/craig/craig-thesis.pdf [November 2016].

Ginsberg, J., Mohebbi, M.H., Patel, R.S., Brammer, L., Smolinski, M.S., and Brilliant, L. (2009). Detecting influenza epidemics using search engine query data. *Nature, 457*(7232), 1012-1014.

Giorgi, E., Sesay, S.S., Terlouw, D.J., and Diggle, P.J. (2015). Combining data from multiple spatially referenced prevalence surveys using generalized linear geostatistical models. *Journal of the Royal Statistical Society: Series A (Statistics in Society), 178*(2), 445-464.

Goel, S., Hofman, J.M., Lahaie, S., Pennock, D.M., and Watts, D.J. (2010). Predicting consumer behavior with web search. *Proceedings of the National Academy of Sciences, 107*(41), 17486-17490.

Goerge, R.M., Smithgall, C., Seshadri, R., and Ballard, P. (2010). *Illinois Families and Their Use of Multiple Service Systems*. Chicago, IL: Chapin Hall. Available: https://www.chapinhall.org/sites/default/files/publications/Multiple%20Systems_IB_03_01_10_0.pdf [November 2016].

Goldreich, O., Micali, S., and Wigderson, A. (1987). How to play any mental game. In *Proceedings of the 19th AMC Symposium on Theory of Computing* (pp. 218-229). New York: ACM Publishing. Available: http://www.math.ias.edu/~avi/PUBLICATIONS/MYPAPERS/GMW87/GMW87.pdf [November 2016].

Grady, S., Bielick, S., and Aud, S. (2010). *Trends in the Use of School Choice: 1993 to 2007*. NCES 2010-004. Washington, DC: National Center for Education Statistics, Institute of Education Sciences, U.S. Department of Education. Available: http://nces.ed.gov/pubs2010/2010004.pdf [November 2016].

Groves, R. (2013). Improving government, academic and industry data-sharing opportunities. In J.A. Krosnick, S. Presser, K. Husbands Fealing, and S. Ruggles (Eds.), *The Future of Survey Research: Challenges and Opportunities*. Available: https://www.nsf.gov/sbe/AC_Materials/The_Future_of_Survey_Research.pdf [November 2016].

Habermann, H. (2010). Future of innovation in the federal statistical system. *The ANNALS of the American Academy of Political and Social Science, 631*(1), 194-203.

Haney, S., Machanavajjhala, A., Abowd, J., Graham, M., Kutzbach, M., and Vilhuber, L. (2017, Forthcoming). *Utility Cost of Formal Privacy for Releasing National Employer-Employee Statistics*. SIGMOD 2017, Raleigh, NC.

Hansen, M., Hurwitz, W.N., and Madow, W.G. (1953a). *Sample Survey Methods and Theory, Volume 1*. New York: John Wiley & Sons.

Hansen, M., Hurwitz, W.N., and Madow, W.G. (1953b). *Sample Survey Methods and Theory, Volume 2*. New York: John Wiley & Sons.

Hansen, M.H., Hurwitz, W.N., Nisselson, H., and Sternberg, J. (1955). The redesign of the Current Population Survey. *Journal of the American Statistical Association, 50*, 701-719.

Hartman, K., Habermann, H., Harris-Kojetin, B., Jones, C., Louis, T., and Gelman, A. (2014). Strength under pressure/A world without statistics. *Significance, 11*(4), 44-47.

Herzog, T.N., Scheuren, F.J., and Winkler, W.E. (2007). *Data Quality and Record Linkage Techniques*. New York: Springer.

Hoff, N.G. (1981). Overview of the consumer expenditure surveys. *Advances in Consumer Research, 8*, 245-250.

Holdren, J.P. (2010). Social science data and the shaping of national policy. *The ANNALS of the American Academy of Political and Social Science, 631*(1), 18-21.

Holt, D.T. (2007). The official statistics Olympic challenge: Wider, deeper, quicker, better, cheaper. *The American Statistician, 61*(1), 1-8.

Homer, N., Szelinger, S., Redman, M., Duggan, D., Tembe, W., Muehling, J., Pearson, J., Stephan, D., Nelson, S., and Craig, D. (2008). Resolving individuals contributing trace amounts of DNA to highly complex mixtures using high-density SNP genotyping microarrays. *PLoS Genetics, 4*(8), e1000167. Available: http://journals.plos.org/plosgenetics/article?id=10.1371/journal.pgen.1000167 [November 2016].

Horrigan, M. (2013). *Big Data and Official Statistics.* Available: https://www.bls.gov/osmr/symp2013_horrigan.pdf [November 2016].

Hoyakem, C., Bollingerm, C., and Ziliak, J. (2014). *The Role of CPS Nonresponse on the Level and Trend in Poverty.* UKCPR Discussion Paper Series, DP 2014-05. Available: http://www.ukcpr.org/sites/www.ukcpr.org/files/documents/DP2014-05_0.pdf [November 2016].

Infas. (2010). *Der Überwachte Bürger Zwischen Apathie Und Protest—Erste Ergebnisse.* Available: http://www.vorratsdatenspeicherung.de/images/infas-umfrage.pdf [November 2016].

Institute of Medicine. (2009). *Beyond the HIPAA Privacy Rule: Enhancing Privacy, Improving Health Through Research.* S.J. Nass, L.A. Levit, and L.O. Gostin (Eds.). Board on Health Sciences Policy, Board on Health Care Services, Committee on Health Research and the Privacy of Health Information: The HIPAA Privacy Rule. Washington, DC: The National Academies Press.

Kasiviswanathan, S.P., Rudelson, M., and Smith, A. (2013). *The Power of Linear Reconstruction Attacks.* Proceedings of the 45th Annual ACM Symposium on the Theory of Computing, Palo Alto, CA, June 2-4. Available: https://arxiv.org/pdf/1210.2381v1.pdf [November 2016].

Kliss, B., and Scheuren, F.J. (1978). The 1973 CPS-IRS-SSA Exact Match Study. *Social Security Bulletin, 51*(7), 23-31.

Kraus, R. (2013). Statistical déjà vu: The National Data Center Proposal of 1965 and its descendants. *Journal of Privacy and Confidentiality, 5*(1), 1-37.

Lavallée, P. (2000). *Combining Survey and Administrative Data: Discussion Paper.* Proceedings of the Second International Conference on Establishment Surveys, Survey Methods for Businesses, Farms, and Institutions, Buffalo, NY.

Lazer, D., Kennedy, R., King, G., and Vespignani, A. (2014). The parable of Google Flu: Traps in big data analysis. *Science, 343*(6176), 1203-1205.

Lee, H., Warren, A., and Gill, L. (2015). *Cheaper, Faster, Better: Are State Administrative Data the Answer?* OPRE Report 2015-09. Available: http://www.acf.hhs.gov/sites/default/files/opre/mihope_strongstart_2yr_2015.pdf [November 2016].

Lohr, S.L., and Raghunathan, T.E. (in press). Combining survey data with other data sources. *Statistical Science.*

Louis, T.A. (2016). *Discussion of Combining Information from Survey and non-Survey Data Sources: Challenges and Opportunities by Sharon Lohr and Trivellore Raghunathan.* Proceedings of the 130th CNSTAT Meeting Public Seminar, Washington, DC, May 6. Available: http://sites.nationalacademies.org/cs/groups/dbassesite/documents/webpage/dbasse_172505.pdf [December 2016].

Lum, K., and Isaac, W. (2016). To predict and serve? *Significance, 13*, 14-19.

Machanavajjhala, A., Kifer, D., Abowd, J., Gehrke, J., and Vilhuber, L. (2008). *Privacy: Theory Meets Practice on the Map*. Proceedings of the IEEE 24th International Conference on Data Engineering, Cancun, Mexico, April 7-12. Available: http://www.cse.psu.edu/~duk17/papers/PrivacyOnTheMap.pdf [November 2016].

Malone, S., and Mutikani, L. (2012). Jack Welch sets Twitter ablaze with Obama job jab. *Chicago Tribune*, October 5. Available: http://articles.chicagotribune.com/2012-10-05/news/sns-rt-us-usa-economy-jackwelchbre8941cr-20121005_1_tweet-jack-welch-alan-krueger [November 2016].

Manyika, J., Chui, M., Brown, B., Bughin, J., Dobbs, R., Roxburgh, C., and Byers, A.H. (2011). Big data: The next frontier for innovation, competition, and productivity. *McKinsey Global Institute,* May. Available: http://www.mckinsey.com/business-functions/digital-mckinsey/our-insights/big-data-the-next-frontier-for-innovation [November 2016].

Manzi, G., Spiegelhalter, D.J., Turner, R.M., Flowers, J., and Thompson, S.G. (2011). Modelling bias in combining small area prevalence estimates from multiple surveys. *Journal of the Royal Statistical Society, 174*(1), 31-50.

Marchetti, S., Giusti, C., Pratesi, M., Salvati, N., Giannotti, F., Pedreschi, D., Rinzivillo, S., Pappalardo, L., and Gabrielli, L. (2015). Small area model-based estimators using big data sources. *Journal of Official Statistics, 31*(2).

McPhee, C., Bielick, S., Masterton, M., Flores, L., Parmer, R., Amchin, S., Stern, S., and McGowan, H. (2015). *National Household Education Surveys Program of 2012: Data File User's Manual.* NCES 2015-030. Washington, DC: National Center for Education Statistics, Institute of Education Sciences, U.S. Department of Education. Available: https://nces.ed.gov/nhes/pdf/userman/NHES_2012_UsersManual.pdf [November 2016].

Meyer, B.D., Mok, W.K.C., and Sullivan, J.X. (2015). Household surveys in crisis. *Journal of Economic Perspectives, 29*(4), 199-226.

Miller, P.V. (2010). Presidential address: The road to transparency in survey research. *Public Opinion Quarterly, 74*(3), 602-606.

Muthukrishnan, S., and Nikolov, A. (2012). Optimal private halfspace counting via discrepancy. In *Proceedings of the Forty-Fourth Annual ACM Symposium on Theory of Computing* (pp. 1285-1292). New York: ACM Publishing. Available: http://dl.acm.org/citation.cfm?id=2214090&dl=ACM&coll=DL&CFID=698830182&CFTOKEN=71109411 [November 2016].

Narayanan, A., and Shmatikov, V. (2008). *Robust De-Anonymization of Large Sparse Datasets (How to Break Anonymity of the Netflix Prize Dataset)*. Proceedings of the 29th IEEE Symposium on Security and Privacy, Oakland, CA, May 18-21. Available: https://www.cs.cornell.edu/~shmat/shmat_oak08netflix.pdf [November 2016].

National Academies of Sciences, Engineering, and Medicine. (2016). *Reducing Response Burden in the American Community Survey: Proceedings of a Workshop*. Committee on National Statistics, Division of Behavioral and Social Sciences and Education. Washington, DC: The National Academies Press.

National Institutes of Health. (2016). *National Institutes of Health Fiscal Year 2017 Budget Request*. Available: https://officeofbudget.od.nih.gov/pdfs/FY17/31-Overview.pdf [November 2016].

National Research Council. (1979). *Privacy and Confidentiality as Factors in Survey Response.* Panel on Privacy and Confidentiality as Factors in Survey Response, Committee on National Statistics, Division of Behavioral and Social Sciences and Education. Washington, DC: National Academy Press.

National Research Council. (1993). *Private Lives and Public Policies: Confidentiality and Accessibility of Government Statistics.* G.T. Duncan, T.B. Jabine, and V.A. De Wolf (Eds.). Panel on Confidentiality and Data Access, Committee on National Statistics, Division of Behavioral and Social Sciences and Education. Washington, DC: National Academy Press.

National Research Council. (2003). *Statistical Issues in Allocating Funds by Formula.* T.A. Louis, T.B. Jabine, and M.A. Gerstein (Eds.). Panel on Formula Allocations, Committee on National Statistics, Division of Behavioral and Social Sciences and Education. Washington, DC: The National Academies Press.

National Research Council. (2004). *Reengineering the 2010 Census: Risks and Challenges.* D.L. Cork, M.L. Cohen, and B.F. King (Eds.). Panel on Research on Future Census Methods, Committee on National Statistics, Division of Behavioral and Social Sciences and Education. Washington, DC: The National Academies Press.

National Research Council. (2005). *Expanding Access to Research Data: Reconciling Risks and Opportunities.* Panel on Data Access for Research Purposes, Committee on National Statistics, Division of Behavioral and Social Sciences and Education. Washington, DC: The National Academies Press.

National Research Council. (2007). *Understanding Business Dynamics: An Integrated Data System for America's Future.* J. Haltiwanger, L.M. Lynch, and C. Mackie (Eds.). Panel on Measuring Business Formation, Dynamics, and Performance; Committee on National Statistics; Division of Behavioral and Social Sciences and Education. Washington, DC: The National Academies Press.

National Research Council. (2008). *Using the American Community Survey for the National Science Foundation's Science and Engineering Workforce Statistics Programs.* Panel on Assessing the Benefits of the American Community Survey for the NSF Division of Science Resources Statistics, Committee on National Statistics, Division of Behavioral and Social Sciences and Education. Washington, DC: The National Academies Press.

National Research Council. (2009a). *Ensuring the Quality, Credibility, and Relevance of U.S. Justice Statistics.* R.M. Groves and D.L. Cork (Eds.). Panel to Review the Programs of the Bureau of Justice Statistics, Committee on National Statistics, Division of Behavioral and Social Sciences and Education. Washington, DC: The National Academies Press.

National Research Council. (2009b). *Reengineering the Survey of Income and Program Participation.* C.F. Citro and J.K. Scholz (Eds.). Panel on the Census Bureau's Reengineered Survey of Income and Participation, Committee on National Statistics, Division of Behavioral and Social Sciences and Education. Washington, DC: The National Academies Press.

National Research Council. (2010a). *Data on Federal Research and Development Investments: A Pathway to Modernization.* Panel on Modernizing the Infrastructure of the National Science Foundation Federal Funds Survey, Committee on National Statistics, Division of Behavioral and Social Sciences and Education. Washington, DC: The National Academies Press.

National Research Council. (2010b). *Envisioning the 2020 Census.* L.D. Brown, M.L. Cohen, D.L. Cork, and C.F. Citro (Eds.). Panel on the Design of the 2020 Census Program of Evaluations and Experiments, Committee on National Statistics, Division of Behavioral and Social Sciences and Education. Washington, DC: The National Academies Press.

National Research Council. (2011). *Change and the 2020 Census: Not Whether but How.* Thomas M. Cook, Janet L. Norwood, and Daniel L. Cork (Eds.). Panel to Review the 2010 Census, Committee on National Statistics, Division of Behavioral and Social Sciences and Education. Washington, DC: The National Academies Press.

National Research Council. (2013a). *Nonresponse in Social Science Surveys: A Research Agenda.* R. Tourangeau and T.J. Plewes (Eds.). Panel on a Research Agenda for the Future of Social Science Data Collection, Committee on National Statistics, Division of Behavioral and Social Sciences and Education. Washington, DC: The National Academies Press.

National Research Council. (2013b). *Principles and Practices for a Federal Statistical Agency.* Committee on National Statistics, Division of Behavioral and Social Sciences and Education. Washington, DC: The National Academies Press.

National Research Council. (2014a). *Capturing Change in Science, Technology, and Innovation: Improving Indicators to Inform Policy.* R.E. Litan, A.W. Wyckoff, and K.H. Fealing (Eds.). Panel on Developing Science, Technology, and Innovation Indicators for the Future; Board on Science, Technology, and Economic Policy; Division of Policy and Global Affairs. Washington, DC: The National Academies Press.

National Research Council. (2014b). *Civic Engagement and Social Cohesion: Measuring Dimensions of Social Capital to Inform Policy.* K. Prewitt, C.D. Mackie, and H. Habermann (Eds.). Panel on Measuring Social and Civic Engagement and Social Cohesion in Surveys, Committee on National Statistics, Division of Behavioral and Social Sciences and Education. Washington, DC: The National Academies Press.

National Research Council. (2015). *Realizing the Potential of the American Community Survey: Challenges, Tradeoffs, and Opportunities.* Panel on Addressing Priority Technical Issues for the Next Decade of the American Community Survey, Committee on National Statistics, Division of Behavioral and Social Sciences and Education. Washington, DC: The National Academies Press.

Norwood, J. (1995). *Organizing to Count: Change in the Federal Statistical System.* Washington, DC: Urban Institute Press.

Norwood, J. (2016). Politics and federal statistics. *Statistics and Public Policy, 3*(1), 1-8.

Parten, M. (1950). *Surveys, Polls, and Samples: Practical Procedures.* New York: Harper & Brothers.

Polivka, A.E., and Miller, S.M. (1995). *The CPS After the Redesign: Refocusing the Economic Lens.* Available: http://www.bls.gov/osmr/pdf/ec950090.pdf [November 2016].

Prell, M., Bradsher-Fredrick, H., Comisarow, C., Cornman, S., Cox, C., Denbaly, M., Martinez, R.W., Sabol, W., and Vile, M. (2009). *Profiles in Success of Statistical Uses of Administrative Data.* Available: http://www.bls.gov/osmr/fcsm.pdf [November 2016].

President's Council of Advisors on Science and Technology. (2014). *Big Data and Privacy: A Technological Perspective.* Available: https://obamawhitehouse.archives.gov/sites/default/files/microsites/ostp/PCAST/pcast_big_data_and_privacy_-_may_2014.pdf [February 2017].

Prewitt, K. (2010). Science starts not after measurement, but with measurement. *The ANNALS of the American Academy of Political and Social Science, 631*(1), 7-16.

Puts, M.J.H., Tennekes, M., Daas, P.J.H., and Blois, C.D. (2016). *Using Huge Amounts of Road Sensor Data for Official Statistics.* Proceedings of the European Conference on Quality in Official Statistics (Q2016), Madrid, Spain. Available: http://www.pietdaas.nl/beta/pubs/pubs/q2016Final00177.pdf [November 2016].

Ramzy, A. (2016). Australia stops online collection of census data after cyberattacks. *The New York Times*, August 10. Available: http://www.nytimes.com/2016/08/11/world/australia/census-cyber-attack.html [November 2016].

Rand, M., and Catalano, S.M. (2007). *Crime Victimization, 2006.* Bureau of Justice Statistics, NCJ 219413. Available: http://www.bjs.gov/content/pub/pdf/cv06.pdf [November 2016].

Reamer, A. (2014). *Stumbling into the Great Recession: How and Why GDP Estimates Kept Economists and Policymakers in the Dark*. Washington, DC: The George Washington Institute of Public Policy. Available: https://bea.gov/about/pdf/Reamer%20GDP%20Research%20Note%2004-25-14%20(1).pdf [November 2016].

Reamer, A., and Carpenter, R.B. (2010). Surveying for dollars: The role of the American Community Survey in the geographic distribution of federal funds. *Brookings*, July 26. Available: https://www.brookings.edu/research/surveying-for-dollars-the-role-of-the-american-community-survey-in-the-geographic-distribution-of-federal-funds [December 2016].

Roberts, D. (1997). *Implementing the National Incident Based Reporting System: A Status Report*. Bureau of Justice Statistics, NCJ 165581. Available: https://www.bjs.gov/content/pub/pdf/INIBRS.pdf [November 2016].

Robin, N., Klein, T., and Jütting, J. (2016). *Public-Private Partnerships for Statistics: Lessons Learned, Future Steps: A Focus on the Use of Non-Official Data Sources for National Statistics and Public Policy*. PARIS21, OECD Development Co-operation Working Papers, No. 27. Available: http://www.oecd-ilibrary.org/development/public-private-partnerships-for-statistics-lessons-learned-future-steps_5jm3nqp1g8wf-en [December 2016].

Rotenberg, M. (2000). *Preserving Privacy in the Information Society*. Available: http://www.unesco.org/webworld/infoethics_2/eng/papers/paper_10.htm [November 2016].

Sahai, A., and Waters, B. (2005). *Fuzzy Identity-Based Encryption*. Proceedings of Eurocrypt 2005, Aarhus, Denmark, May 22-26. Available: https://eprint.iacr.org/2004/086.pdf [December 2016].

Saslow, E. (2012). "Jobs Day": Monthly release of employment data an economic, political obsession. *Washington Post*, March 9. Available: https://www.washingtonpost.com/national/jobs-day-an-economic-and-political-obsession/2012/03/09/gIQADZPW1R_story.html [November 2016].

Schenker, N., and Raghunathan, T.E. (2007). Combining information from multiple surveys to enhance estimation of measures of health. *Statistics in Medicine, 26*(8), 1802-1811.

Schulte Nordholt, E. (2014). Introduction to the Dutch Census 2011. In *Dutch Census 2011: Analysis and Methodology* (pp. 7-18). The Hague/Heerlen: Statistics Netherlands. Available: https://www.cbs.nl/NR/rdonlyres/5FDCE1B4-0654-45DA-8D7E-807A0213DE66/0/2014b57pub.pdf [December 2016].

Seeskin, Z.H., and Spencer, B.D. (2015). *Effect of Census Accuracy on Appointment of Congress and Allocations of Federal Funds*. WP-15-05. Evanston, IL: Institute for Policy Research, Northwestern University. Available: http://www.ipr.northwestern.edu/publications/docs/workingpapers/2015/IPR-WP-15-05.pdf [November 2016].

Singer, E. (2003). The eleventh Morris Hansen lecture: Public perceptions of confidentiality. *Journal of Official Statistics, 19*(4), 333-341.

Singer, E., and Couper, M.P. (2010). Communicating disclosure risk in informed consent statements. *Journal of Empirical Research on Human Research Ethics, 5*(3), 1-8.

Singer, E., Hoewyk, J.V., and Neugebauer, R.J. (2003). Attitudes and behavior: The impact of privacy and confidentiality concerns on participation in the 2000 Census. *Public Opinion Quarterly, 67*(3), 368-384.

Solove, D.J., and Schwartz, P.M. (2015). *Information Privacy Law* (5th Edition). New York: Aspen Publishers.

Statistics Canada. (2016a). *Creating a Modern Framework for an Independent National Statistics Office*. Available: https://www.scribd.com/document/319364233/Statistics-Canada-recommendations-on-new-powers [November 2016].

Statistics Canada. (2016b). *Model-based Principal Field Crop Estimates*. Available: http://www23.statcan.gc.ca/imdb/p2SV.pl?Function=getSurvey&SDDS=5225#a1 [November 2016].

Statistics Denmark. (2014). *The Danish System for Access to Micro Data*. Available: http://www.dst.dk/ext/645846915/0/forskning/Access-to-micro-data-at-Statistics-Denmark_2014.pdf [November 2016].

Struijs, P., and Daas, P. (2014). *Quality Approaches to Big Data in Official Statistics*. Proceedings for the European Conference on Quality in Official Statistics, Vienna, Austria, June 2-5. Available: http://www.pietdaas.nl/beta/pubs/pubs/Q2014_session_33_paper.pdf [November 2016].

Sweeney, L. (1997). Weaving technology and policy together to maintain confidentiality. *The Journal of Law, Medicine & Ethics, 25*(2-3), 98-110.

Sylvester, D., and Lohr, S. (2005). The security of our secrets: A history of privacy and confidentiality in law and statistical practice. *Denver University Law Review, 83*, 147-208. Available: http://www.law.du.edu/images/uploads/denver-university-law-review/v83_i1_sylvesterlohr.pdf [November 2016].

Tran, M. (2015). *FBI Chief: "Unacceptable" That Guardian Has Better Data on Police Violence*. Available: https://www.theguardian.com/us-news/2015/oct/08/fbi-chief-says-ridiculous-guardian-washington-post-better-information-police-shootings [November 2016].

Trépanier, J., Pignal, J., and Royce, D. (2013). *Administrative Data Initiatives at Statistics Canada*. Proceedings for the Federal Committee on Statistical Methodology Research Conference, Washington DC, November 4-6. Available: https://fcsm.sites.usa.gov/files/2014/05/G1_Trepanier_2013FCSM.pdf [November 2016].

Turn, R., and Ware, W.H. (1976). *Privacy and Security Issues in Information Systems*. Available: https://www.rand.org/content/dam/rand/pubs/papers/2008/P5684.pdf [November 2016].

U.N. Economic Commission for Europe. (2011). *Using Administrative and Secondary Sources for Official Statistics: A Handbook of Principles and Practices*. Available: http://www1.unece.org/stat/platform/display/adso/Using+Administrative+and+Secondary+Sources+for+Official+Statistics [November 2016].

U.N. Economic and Social Council. (2014). *Report of the Global Group on Big Data for Official Statistics*. Available: http://unstats.un.org/unsd/statcom/doc15/2015-4-BigData-E.pdf [November 2016].

U.N. Economic and Social Council. (2016). *Report of the Global Working Group on Big Data for Official Statistics*. Available: http://unstats.un.org/unsd/statcom/47th-session/documents/2016-6-Big-data-for-official-statistics-E.pdf [November 2016].

U.N. Global Pulse. (2014). *Nowcasting Food Prices in Indonesia Using Social Media Signals*. Global Pulse Project Series No. 1. Available: http://www.unglobalpulse.org/sites/default/files/UNGP_ProjectSeries_Nowcasting_Food_Prices_2014.pdf [November 2016].

U.N. Global Pulse. (2016). *Postal Network's Global Big Data Can Be Key to Understanding Nations' Wellbeing*. Available: http://unglobalpulse.org/news/postal-big-data-key-to-understanding-wellbeing [November 2016].

U.S. Census Bureau. (2015). *U.S. Census Bureau's Budget Estimates: Fiscal Year 2016*. Available: http://www.osec.doc.gov/bmi/budget/FY16CJ/Census_2016_CJ.pdf [November 2016].

U.S. Department of Commerce. (2014). *Fostering Innovation, Creating Jobs, Driving Better Decisions: The Value of Government Data*. Economics and Statistics Administration. Available: http://esa.gov/sites/default/files/revisedfosteringinnovationcreatingjobsdrivingbetterdecisions-thevalueofgovernmentdata.pdf [November 2016].

U.S. Department of Health, Education, and Welfare. (1973). *Records, Computers, and the Rights of Citizens*. Available: https://www.justice.gov/opcl/docs/rec-com-rights.pdf [December 2016].

U.S. Department of the Treasury. (2014). *General Explanations of the Administration's Fiscal Year 2015 Revenue Proposals*. Available: https://www.treasury.gov/resource-center/tax-policy/Documents/General-Explanations-FY2015.pdf [November 2016].

U.S. Government Accountability Office. (2003). *Formula Grants: 2000 Census Redistributes Federal Funding among States*. GAO-03-178. Available: http://www.gao.gov/products/GAO-03-178 [November 2016].

U.S. Government Accountability Office. (2006). *Federal Information Collection: A Reexamination of the Portfolio of Major Federal Household Surveys Is Needed*. GAO-07-62. Available: http://www.gao.gov/products/GAO-07-62 [November 2016].

U.S. Government Accountability Office. (2009a). *Formula Grants: Census Data Are among Several Factors That Can Affect Funding Allocations*. GAO-09-832T. Available: http://www.gao.gov/products/GAO-09-832T [November 2016].

U.S. Government Accountability Office. (2009b). *Funding for the Largest Federal Assistance Programs Is Based on Census-Related Data and Other Factors*. GAO-10-263. Available: http://www.gao.gov/new.items/d10263.pdf [November 2016].

U.S. Office of Management and Budget. (1985). *Statistical Policy Directive No. 3: Compilation, Release, and Evaluation of Principal Federal Economic Indicators*. Available: https://obamawhitehouse.archives.gov/sites/default/files/omb/assets/omb/inforeg/statpolicy/dir_3_fr_09251985.pdf [February 2017].

U.S. Office of Management and Budget. (2006). *Statistical Policy Directive No. 2: Standards and Guidelines for Statistical Surveys*. Available: https://obamawhitehouse.archives.gov/sites/default/files/omb/inforeg/statpolicy/standards_stat_surveys.pdf [February 2017].

U.S. Office of Management and Budget. (2007). *Implementation Guidance for the Title V of the E-Government Act, Confidential Information Protection and Statistical Efficiency Act of 2002 (CIPSEA)*. Available: https://obamawhitehouse.archives.gov/sites/default/files/omb/assets/omb/fedreg/2007/061507_cipsea_guidance.pdf [February 2017].

U.S. Office of Management and Budget. (2008). *Statistical Policy Directive No. 4: Release and Dissemination of Statistical Products Produced by Federal Statistical Agencies*. Available: https://obamawhitehouse.archives.gov/sites/default/files/omb/fedreg/2008/030708_directive-4.pdf [February 2017].

U.S. Office of Management and Budget. (2014a). *M-14-06: Guidance for Providing and Using Administrative Data for Statistical Purposes*. Available: https://obamawhitehouse.archives.gov/sites/default/files/omb/memoranda/2014/m-14-06.pdf [February 2017].

U.S. Office of Management and Budget. (2014b). *Statistical Policy Directive No. 1: Fundamental Responsibilities of Federal Statistical Agencies and Recognized Statistical Units*. Available: https://www.gpo.gov/fdsys/pkg/FR-2014-12-02/pdf/2014-28326.pdf [November 2016].

U.S. Office of Management and Budget. (2015a). Chapter 7: Building evidence with administrative data. In *Analytical Perspectives: Budget of the United States Government: Fiscal Year 2016* (pp. 65-73). Washington, DC: Government Printing Office. Available: https://www.gpo.gov/fdsys/pkg/BUDGET-2016-PER/pdf/BUDGET-2016-PER-4-3.pdf [December 2016].

U.S. Office of Management and Budget. (2015b). *Statistical Programs of the United States Government*. Available: https://obamawhitehouse.archives.gov/sites/default/files/omb/assets/information_and_regulatory_affairs/statistical-programs-2016.pdf [February 2017].

U.S. Office of Management and Budget. (2016). Chapter 7: Building the capacity to produce and use evidence. In *Analytical Perspectives: Budget of the United States Government: Fiscal Year 2017* (pp. 69-77). Washington, DC: Government Printing Office. Available: https://obamawhitehouse.archives.gov/sites/default/files/omb/budget/fy2017/assets/ap_7_evidence.pdf [February 2017].

van Tuinen, H.K. (2009). Innovative statistics to improve our notion of reality. *Journal of Official Statistics, 25*(4), 431-465.

Wagner, D., and Layne, M. (2014). *The Person Identification Validation System (PVS): Applying the Center for Administrative Records Research and Applications' (CARRA) Record Linkage Software*. Working Paper No. 2014-01. Available: https://www.census.gov/srd/carra/CARRA_PVS_Record_Linkage.pdf [November 2016].

Wallman, K.K., and Harris-Kojetin, B.A. (2004). Implementing the Confidential Information Protection and Statistical Efficiency Act of 2002. *Chance, 17,* 21-25.

Warren, J. (2016). *Privacy Furore as Australians Prepare for Census*. Available: http://www.forbes.com/sites/justinwarren/2016/08/04/privacy-furore-as-australians-prepare-for-census/#1378c7c7171c [November 2016].

Yang, S., Santillana, M., and Kou, S.C. (2015). Accurate estimation of influenza epidemics using Google search data via ARGO. *Proceedings of the National Academy of Sciences, 112*(47), 14473-14478. Available: www.pnas.org/cgi/doi/10.1073/pnas.1515373112 [December 2016].

Yates, F. (1949). *Sampling Methods for Censuses and Surveys*. London, UK: Charles Griffin and Company.

Zolas, N., Goldschlag, N., Jarmin, R., Stephan, P., Owen-Smith, J., Rosen, R.F., Allen, B.M., Weinberg, B.A., and Lane, J.I. (2015). Wrapping it up in a person: Examining employment and earnings outcomes for Ph.D. recipients. *Science, 350*(6266), 1367-1371.

Zukerberg, A. (2010). *Redesigning the National Household Education Survey (NHES)*. Available: http://www.bls.gov/cex/aaporsrvyredesign2010zuckerb1.pdf [November 2016].

Appendix A

Workshop Agendas

FIRST MEETING OF THE PANEL ON IMPROVING FEDERAL STATISTICS FOR POLICY AND SOCIAL SCIENCE RESEARCH USING MULTIPLE DATA SOURCES AND STATE-OF-THE-ART ESTIMATION METHODS

Friday, September 4, 2015
The National Academies of Sciences, Engineering, and Medicine
2101 Constitution Avenue, NW, NAS Members' Room
Washington, DC

9:00–9:30 a.m	**Welcome and Introductions** Robert M. Groves, *Chair* Connie Citro, *Director, Committee on National Statistics* Stuart Buck, *Laura and John Arnold Foundation*
9:30–10:30 a.m.	**Discussion of Statement of Task** Quick take on purpose and optimal orientation of the study from panel members (5 minutes each)
10:30–10:45 a.m.	**Break**
10:45 a.m.–12:15 p.m.	**Dialogue with Federal Statistical System Stakeholders** Brian Moyer, Bureau of Economic Analysis Bill Sabol, Bureau of Justice Statistics

Erica Groshen, Bureau of Labor Statistics
John Thompson, U.S. Census Bureau
Adam Siemenski, Energy Information Administration
Joe Reilly, National Agricultural Statistics Service
Charlie Rothwell, National Center for Health Statistics
John Gawalt, National Center for Science and Engineering Statistics
Katherine Wallman, Office of Management and Budget

12:15–1:15 p.m. **Working Lunch** (available in room)

1:15 p.m. **Adjournment of Public Session**

AGENDA FOR WORKSHOP ON THE USE OF ALTERNATIVE AND MULTIPLE DATA SOURCES FOR FEDERAL STATISTICS

December 16, 2015
The National Academies of Sciences, Engineering, and Medicine
2101 Constitution Avenue, NW, NAS Lecture Room
Washington, DC

8:45 a.m. **Welcome, Introductions, and Goals of the Workshop**
Robert Groves, Chair

9:00–10:15 a.m. **Session I: Legal and Policy Issues of Using Alternative Data Sources**

9:00 a.m. *Legal and Policy Issues for Statistical Agencies Accessing and Providing Administrative Data: An Overview*
Shelly Martinez, OMB

9:20 a.m. *Accessing Government Records: Authorities and Agreements*
Amy O'Hara, U.S. Census Bureau

9:40 a.m. *Public Attitudes toward Possible Use of Administrative Records to Supplement the U.S. 2020 Census*
Jennifer Hunter Childs, U.S. Census Bureau

10:00 a.m. Question & Answer

10:15–10:30 a.m. **Break**

10:30 a.m.–12:00 p.m.	**Session II: Evaluation of Errors in Alternative Sources**
10:30 a.m.	*Management of Quality, Cost and Risk in BLS Work with Alternative Data Sources* John Eltinge, BLS
10:50 a.m.	*Matching American Housing Survey to tax assessment data: some preliminary findings* Shawn Bucholtz, HUD
11:05 a.m.	*Results from the 2012 Commercial Buildings Energy Consumption Survey (CBECS) Energy Usage Data Validation Study* Joelle Michaels, EIA
11:20 a.m.	*Using Proprietary Household and Retail Scanner Data in Food Policy Research* Megan Sweitzer, ERS
11:35 a.m.	An Automated System for Transforming National Criminal History Records into Statistical Databases Howard Snyder, BJS
11:50 a.m.	Question & Answer
12:00–1:00 p.m.	**Working Lunch** (box lunches provided for all attendees)
1:00–3:15 p.m.	**Session III: Combining Multiple Data Sources**
1:00 p.m.	*Using Commercial Data to Estimate Spending* Dennis Fixler, BEA
1:30 p.m.	*Overview of Census Bureau Efforts to Combine Data to Improve Economic and Social Measurement* Ron Jarmin, Census
2:00 p.m.	*On Combining Multiple Sources of Information to Enhance NASS Crop Estimates* Nathan Cruze, NASS

2:15 p.m.	*Integration of Administrative Record Data into Sample Surveys Conducted by the National Center for Education Statistics* Chris Chapman, NCES
2:30 p.m.	*Using Public And Private Data Sets to Enhance the Research Value of FoodAPS* John Kirlin, ERS
2:45 p.m.	*Measuring All Freight Movement in the United States* Rolf Schmitt, BTS
3:00 p.m.	Question & Answer
3:15–3:30 p.m.	**Break**
3:30–5:00 p.m.	**Session IV: Examining Alternative Data Sources**
3:30 p.m.	*Use of Electronic Health Records by the National Health Care Surveys* Clarice Brown, NCHS
3:45 p.m.	*The 2015 Arrest-Related Deaths (ARD) Pilot Program* Mike Planty, BJS
4:00 p.m.	*A Feasibility Study of Linking SED to UMETRICS and ProQuest* Wan-Ying Chang, NCSES
4:15 p.m.	*Objective Measures of Physical Activity: Considerations for Data Management, Processing, and Public Release* Tala Fakhouri, NCHS
4:30 p.m.	*ERS Use of Food Assistance Administrative Records* Mark Prell, ERS
4:45 p.m.	Question & Answer
5:00 p.m.	**Adjourn**

WORKSHOP ON THE CURRENT PRACTICES OF PRIVATE COMPANIES AND THEIR USE OF BIG DATA AND KEY ISSUES AND CHALLENGES WITH PRIVACY AND CONFIDENTIALITY

February 25, 2016
Paul Brest East, Munger Building 4
Stanford University
Stanford, CA

8:30 a.m. **Welcome and Introductions**
Robert Groves, *Chair*

8:45–10:00 a.m. **Data Collection and Assembly**
Moderator: Ophir Frieder, Georgetown University

Modernizing Statistical Data Collection
Joe Reisinger, Premise

10:00–10:15 a.m. **Break**

10:15–11:30 a.m. **Data Sharing and Access**
Moderator: Robert Groves, Georgetown University

Climate Change Data: Management and Distributed Resources
Dean Williams, Lawrence Livermore National Laboratory

Corporate Data Access and Sharing
Stephen J. Eglash, Stanford Data Science Initiative

11:30 a.m.–
12:30 p.m. **Preserving Privacy**
Moderator: Marc Rotenberg, Electronic Privacy Information Center

The Practice of (Privacy-Preserving?) Data Sharing
Aleksandra Korolova, University of Southern California

12:30–1:30 p.m. **Lunch**

1:30–2:45 p.m. **Transaction Data: Analysis and Access**
Moderator: Roberto Rigobon, MIT Sloan School of Management

	An Overview of MasterCard's SpendingPulse Kamalesh Rao, MasterCard Advisors
	The J.P. Morgan Chase Institute: Challenges, Opportunities and Next Steps in Using Proprietary Transaction Level Data for Economic Research Rachel Pacheco, J.P. Morgan Chase Institute
2:45–4:00 p.m.	**Social Media Data** **Moderator:** Frauke Kreuter, University of Maryland
	Data @ LinkedIn Ya Xu and Kevin Morsony, LinkedIn
4:00–5:00 p.m.	**Combining Survey Data with Organic Data** **Moderator:** Michael Chernew, Harvard Medical School
	Google Tools for Data Hal Varian, Google
5:00 p.m.	**Adjourn**

WORKSHOP ON STATE AND LOCAL GOVERNMENTS' USE OF ALTERNATIVE AND MULTIPLE DATA SOURCES

June 1, 2016
The National Academies of Sciences, Engineering, and Medicine
Keck Center, 500 Fifth Street, NW, Room 100
Washington, DC

8:30 a.m.	**Welcome, Introductions, and Goals of the Workshop** Robert Groves, Chair; Georgetown University
8:35 a.m.	**Session I: Creating and Building Data Systems** **Moderator:** Jim Lynch, University of Maryland
	Federalism Run Amok: Building a Data Infrastructure for K-12 Education in the States Jack Buckley, The College Board
	Implementation of NCS-X/NIBRS in Police Jurisdictions Howard Snyder, Bureau of Justice Statistics

	Census Bureau Experiences in Obtaining State UI and SNAP Data Ron Jarmin, U.S. Census Bureau
10:20 a.m.	**Break**
10:35 a.m.	**Session II: Using Alternative Data Sources** **Moderator:** Ophir Frieder, Georgetown University
	Sensing Cities Steve Koonin, NYU Center for Urban Science and Progress
11:15 a.m.	**Session III: Integrating State Data Systems** **Moderator:** H.V. Jagadish, University of Michigan
	Building State Infrastructure to Effectively Manage, Link, and Use Data Rachel Zinn, Workforce Data Quality Campaign
11:55 a.m.	**Working Lunch** (lunch provided for all attendees)
12:50 p.m.	**Session IV: Integrating Local and National Data Systems** **Moderator:** Colm O'Muircheartaigh, NORC/University of Chicago
	Linking City, County, State, and Federal Datasets Robert Goerge, Chapin Hall at the University of Chicago
	Encouraging National, Public, and Private Data Sharing through the Regional Integrated Transportation Information System *Michael Pack, University of Maryland*
2:05 p.m.	**Session V: Governance of Integrated Data Systems** **Moderator:** Frauke Kreuter, University of Maryland
	Common Governance Models for Integrated Data Systems *Whitney LeBoeuf, University of Pennsylvania*
2:45 p.m.	**Break**

3:00 p.m.	**Session VI: Privacy** **Moderator:** Marc Rotenberg, Electronic Privacy Information Center *Privacy Issues with Sensor Data Collections* Michael Froomkin, University of Miami *Information Sharing and Analytics with Privacy by Design* Jeff Jonas, IBM
4:15 p.m.	**Session VII: Issues with Federal Statistical Agency Gaining Access to Datasets** **Moderator:** Cynthia Dwork, Microsoft Research *Discussions with Private Firms about Sharing Data with Federal Statistical Agencies* *Steve Eglash, Stanford University*
5:00 p.m.	**Adjourn**

Appendix B

Biographical Sketches of Panel Members and Staff

Robert M. Groves (*Chair*) is provost, Gerard Campbell professor in the Department of Mathematics and Statistics, and a professor in the Department of Sociology, all at Georgetown University. His research focuses on the effects of the mode of data collection on responses in sample surveys, the social and political influences on survey participation, the use of adaptive research designs to improve the cost and error properties of statistics, and how public concerns about privacy affect attitudes toward statistical agencies. Previously, he served as director of the U.S. Census Bureau, director of the University of Michigan Survey Research Center, and research professor at the Joint Program in Survey Methodology at the University of Maryland. He is an elected member of the National Academy of Sciences, the National Academy of Medicine, the American Academy of Arts and Sciences, and the International Statistical Institute and an elected fellow of the American Statistical Association. His 1989 book, *Survey Errors and Survey Costs,* was named one of the 50 most influential books in survey research by the American Association of Public Opinion Research. He has a bachelor's degree from Dartmouth College, a master's degrees in statistics and sociology from the University of Michigan, and a doctorate from the University of Michigan.

Michael E. Chernew is a professor of health care policy in the Department of Health Care Policy at Harvard Medical School. He is also a research associate of the National Bureau of Economic Research. His research examines areas related to controlling health care spending growth while maintaining or improving the quality of care, including consumer incentives to align patient cost sharing with clinical value. Related research examines the effects of changes in Medicare Advantage payment rates as well as the causes and consequences of rising health care spending and geographic variation in spending, spending growth, and quality. He is a member of the Medicare Payment Advisory Commission (MedPAC), an independent agency that advises Congress. He is a recipient of the John D. Thompson Prize for Young Investigators given by the Association of University Programs in Public Health and of the Alice S. Hersh Young Investigator Award from the Association of Health Services Research. He has a B.A. from the University of Pennsylvania and a Ph.D. in economics from Stanford University.

Piet Daas is a senior methodologist in the Department of Corporate Services, Information Technology, and Methodology and a data scientist in the Center for Big Data Statistics of Statistics Netherlands. His work focuses on the use of secondary (nonsurvey) data for official statistical purposes, which began with the use of administrative data, and more recently has focused on studies in which Internet and other big data sources are used for official statistics. At Statistics Netherlands he is a member of the big data core team, which oversees all big data activities of production, information technology, research, management and training. He teaches the big data component of the European Master of Official Statistics track at the University of Utrecht, is involved in the big data courses of the European Statistical Training Programme, and is a member of the team organizing DataCamps ("hackatons") at the University of Twente. He is active in various European, United Nations, and U.N. Economic Commission for Europe big data initiatives. He has an M.S. and a Ph.D. in the natural sciences with honors from the University of Nijmegen in the Netherlands.

Cynthia Dwork, on leave from Microsoft Research, is the Gordon McKay professor of computer science at the John A. Paulson School of Engineering and Applied Sciences and a Radcliffe alumnae professor at the Radcliffe Institute for Advanced Study, both at Harvard University. Her work focuses on placing privacy-preserving data analysis on a mathematically rigorous foundation: a cornerstone of this work is differential privacy, a strong privacy guarantee frequently permitting highly accurate data analysis. She also does work in cryptography and distributed computing, including work on the first public-key cryptosystem for which breaking a random instance is as hard as solving the hardest instance of the underlying mathematical problem on combating e-mail spam by requiring a proof of computational effort (the technology that underlies hashcash and bitcoin). She is a recipient of the PET Award for Outstanding Research in Privacy Enhancing Technologies given by Microsoft and of the Edsger W. Dijkstra Prize, awarded jointly by the ACM Symposium on Principles of Distributed Computing of the Association for Computing Machinery and the European Association for Theoretical Computer Science (EATCS) Symposium on Distributed Computing. She is a member of the National Academy of Sciences and the National Academy of Engineering and a fellow of the American Academy of Arts and Sciences. She has a B.S.E. from Princeton University and a Ph.D. from Cornell University.

Ophir Frieder is the Robert L. McDevitt, K.S.G., K.C.H.S. and Catherine H. McDevitt L.C.H.S. chair in computer science and information processing at Georgetown University. He is also a professor of biostatistics, bioinformatics, and biomathematics in the Georgetown University Medical Center and the chief scientific offer for UMBRA Health Corporation. He previously served as chair of the Department of Computer Science at Georgetown University. His research interests focus on scalable information retrieval systems spanning search and retrieval and communications issues in multiple domains, systems that are deployed worldwide in commercial and governmental production environments. He is a fellow of the American Association for the Advancement of Science, the Association for Computing Machinery, the Institute of Electrical and Electronics Engineers, and the National Academy of Inventors.

Brian Harris-Kojetin (*Study Director)* is deputy director of the Committee on National Statistics and served as the study director for this project. Previously, he worked at the U.S. Office of Management and Budget (OMB), where he served as a senior statistician in the Statistical and Science Policy Office. He chaired the Federal Committee on Statistical Methodology and was the lead at OMB on issues related to standards for statistical surveys, survey nonresponse, measurement of race and ethnicity, and confidentiality of statistical data. He also previously was senior project leader of research standards and practices at the Arbitron Company and a research psychologist in the Office of Survey Methods Research in the Bureau of Labor Statistics. He is a fellow of the American Statistical Association. He has a B.A. from the University of Denver and a Ph.D. from the University of Minnesota.

H.V. Jagadish is the Bernard A. Galler collegiate professor of electrical engineering and computer science and distinguished scientist at the Institute for Data Science at the University of Michigan in Ann Arbor. Previously, he was head of the Database Research Department at AT&T Labs in Florham Park, New Jersey. He works widely in information management and holds numerous patents in the field. He is a fellow of the Association for Computing Machinery (ACM), serves on the board of the Computing Research

Association, and was a trustee of the VLDB (very large database) Foundation. He is a recipient of the SIGMOD Contributions Award from ACM and of the David E. Liddle Research Excellence Award from the University of Michigan. He has a B.Tech. from the Indian Institute of Technology, Delhi, and an M.S. and a Ph.D. from Stanford University, all in electrical engineering.

Frauke Kreuter is director of the Joint Program in Survey Methodology at the University of Maryland, professor of statistics and methodology at the University of Mannheim, Germany, and head of the statistical methods group at the German Institute for Employment Research. She is also affiliated with the Maryland Population Research Center and the Institute for Social Research in Michigan. Previously, she held positions at the Institute for Statistics at the Ludwig-Maximilians University in Munich, Germany, and in the Department of Statistics at the University of California, Los Angeles. Her research focuses on nonresponse errors, paradata and responsive designs, record linkage, and, recently, issues of linkage consent and generalizability for nonprobability samples. She is an elected fellow of the American Statistical Association and a recipient of the Gertrude M. Cox Statistics Award from the Washington Statistical Society. She serves on the advisory boards of Statistics Canada, Statistics Sweden, and the U.S. Energy Information Association. She has a B.A. and an M.A. from the University of Mannheim and a Ph.D. from the University of Konstanz in Germany.

Sharon Lohr is a vice president and senior statistician at Westat in Rockville, Maryland. Previously, she was dean's distinguished professor of statistics at Arizona State University. Her research has focused on survey sampling, hierarchical models, small-area estimation, missing data, and design of experiments. She is a fellow of the American Statistical Association and an elected member of the International Statistical Institute. She was the inaugural recipient of the Washington Statistical Society's Gertrude M. Cox Statistics Award for contributions to the practice of statistics and a recipient of the society's Morris Hansen Lecture Award. She was recently selected to present the Deming Lecture at the Joint Statistical Meetings. She has a Ph.D. in statistics from the University of Wisconsin–Madison.

James P. Lynch is professor and chair of the Department of Criminology and Criminal Justice at the University of Maryland. Previously, he served as director of the Bureau of Justice Statistics at the U.S. Department of Justice and was a distinguished professor in the Department of Criminal Justice at John Jay College of the City University of New York. He also previously was a professor in and chair of the Department of Justice, Law and Society at American University. His research focuses on victim surveys, victimization risk, the role of coercion in social control, and crime statistics. He has been vice president of the American Society of Criminology and served on the Committee on Law and Justice Statistics of the American Statistical Association. He has a B.A. from Wesleyan University and an M.A. and a Ph.D. in sociology from the University of Chicago.

Colm A. O'Muircheartaigh is professor and former dean of the Harris School of Public Policy Studies and a senior fellow at NORC, both at the University of Chicago. Previously, he was the first director of the Methodology Institute and a faculty member of the Department of Statistics at the London School of Economics and Political Science. The primary focus of his work is on the design of complex surveys across a wide range of populations and topics and on fundamental issues of data quality, including the impact of errors in responses to survey questions, cognitive aspects of question wording, and latent variable models for nonresponse. He is a fellow of the Royal Statistical Society and the American Statistical Association and an elected member of the International Statistical Institute. He has served as a consultant to a wide range of public and commercial organizations around the world, including OECD and the United Nations. He received his undergraduate education at University College Dublin and his graduate education at the London School of Economics.

Trivellore Raghunathan is director of the Survey Research Center and a research professor at the Institute for Social Research at the University of Michigan, where he is also a professor of biostatistics and an associate director of the Center for Research on Ethnicity, Culture and Health in the School of Public Health. He is also a research professor in the Joint Program in Survey Methodology at the University of Maryland. Previously, he was on the faculty in the Department of Biostatistics at the University of Washington. His research interests are in the analysis of incomplete data, multiple imputation, Bayesian methods, design and analysis of sample surveys, combining information from multiple data sources, small-area estimation, confidentiality and disclosure limitation, longitudinal data analysis, and statistical methods for epidemiology. He has developed SAS-based software for imputing the missing values for a complex dataset. He has a Ph.D. in statistics from Harvard University.

Roberto Rigobon is the Society of Sloan Fellows professor of management and professor of applied economics at the Sloan School of Management at the Massachusetts Institute of Technology. He is also a visiting professor at Instituto de Estudios Superiores de Administración (Institute of Advanced Studies in Administration) in Venezuela and a research associate of the National Bureau of Economic Research. His research has addressed the causes of balance-of-payments crises, financial crises, and the propagation of them across countries. He is currently studying the properties of international pricing practices and how to produce alternative measures of inflation. He is one of the two founding members of the Billion Prices Project, as well as a cofounder of PriceStats. He is a member of the Census Bureau's Scientific Advisory Committee and president of the Latin American and Caribbean Economic Association. He has a B.S. in electrical engineering from Universidad Simon Bolivar (Venezuela), an M.B.A. from Instituto de Estudios Superiores de Administración (Venezuela), and a Ph.D. in economics from the Massachusetts Institute of Technology.

Marc Rotenberg is president of the Electronic Privacy Information Center (EPIC) in Washington, D.C., and teaches information privacy law and open government at Georgetown University Law Center. He has testified before Congress on more than 60 occasions and authored more than 50 amicus briefs on emerging privacy and civil liberties issues. He has served on several national and international advisory panels, including the expert panels on Cryptography Policy and Computer Security for OECD and the Legal Experts on Cyberspace Law for UNESCO. He is a founding board member and former chair of the Public Interest Registry, which manages the .org domain. He is a fellow of the American Bar Foundation, a member of the Council on Foreign Relations, and the recipient of several awards, including the World Technology Award in Law from the World Technology Network. He has an A.B. from Harvard College, a J.D. from Stanford Law School, and an LL.M. in international and comparative law from Georgetown University.

COMMITTEE ON NATIONAL STATISTICS

The Committee on National Statistics was established in 1972 at the National Academies of Sciences, Engineering, and Medicine to improve the statistical methods and information on which public policy decisions are based. The committee carries out studies, workshops, and other activities to foster better measures and fuller understanding of the economy, the environment, public health, crime, education, immigration, poverty, welfare, and other public policy issues. It also evaluates ongoing statistical programs and tracks the statistical policy and coordinating activities of the federal government, serving a unique role at the intersection of statistics and public policy. The committee's work is supported by a consortium of federal agencies through a National Science Foundation grant.